Just a Wor

A light-hearted guide
to medical terms

Bernard J. Freedman

Oxford · New York · Tokyo
OXFORD UNIVERSITY PRESS

Oxford University Press, Walton Street, Oxford OX2 6DP
Oxford New York Toronto
Delhi Bombay Calcutta Madras Karachi
Petaling Jaya Singapore Hong Kong Tokyo
Nairobi Dar es Salaam Cape Town
Melbourne Auckland
and associated companies in
Beirut Berlin Ibadan Nicosia

Oxford is a trade mark of Oxford University Press

Published in the United States
by Oxford University Press, New York

First published 1987
Reprinted (with corrections) 1988

British Library Cataloguing in Publication Data
Freedman, Bernard J.
Just a word, Doctor: a light-hearted guide
to medical terms—
(Oxford medical publications).
1. Medicine—Terminology
I. Title
610'.14 R123
ISBN 0-19-261641-2

Library of Congress Cataloging in Publication Data
Freedman, Bernard J.
Just a word, Doctor.
(Oxford medical publications) (Oxford paperbacks)
Consists of essays originally published in the
British medical journal between 1977 and 1986.
1. Medicine—Terminology. I. British medical journal.
II. Title. III. Series. [DNLM: 1. Nomenclature—
collected works. W 15 F853j]
R123.F69 1987 610'.014 87-5544
ISBN 0-19-261641-2 (pbk.)

Printed in Great Britain by
Richard Clay Ltd, Bungay, Suffolk

To Martin Netsky

Preface

The little essays in this book were first published in the *British Medical Journal* at irregular intervals between 1977 and 1986. They have been updated and, thanks to colleagues, errors have been corrected; several have been rewritten. Foreign words abound and, where these are not written in Roman type in the original, they have been transliterated to conventional equivalents. For the occasional phonetic transcription, where pronunciation is relevant, I have sometimes adopted an idiosyncratic system which I hope will leave no doubt in the mind's ear of the reader.

I have two hats, metaphorically speaking. When I wear my lexicographic hat, I record what people write and say. When I wear my schoolmaster's hat, I allow myself the freedom to opine that a particular word is or is not appropriate, and may express a preference for one word rather than another; only rarely would I say a word is wrongly used in a specific context or wrongly spelt.

There is much about etymology in this collection. Etymology is of historical interest, and the pursuit of origins can be fun. Etymology should never be invoked to alter a long-established spelling, though there are many people who like to do this in the belief that so doing achieves 'accuracy'. Imagine the chaos that would ensue if the spelling of all English words were revised in conformation with their origins!

Just as the articles appeared originally at intervals, so most readers will, I think, enjoy them best by browsing from time to time. But everyone according to his taste. Two of the articles are much longer than average; 'Fingers and digits' was written for a Christmas edition of the *British Medical Journal*, and 'Caucasian' was commissioned.

London B. J. F.
1986

Acknowledgements

Anyone who attempts to cover an entire vocabulary, even of a single discipline, is heavily dependent on advice from experts. I am indebted to Dr J. A. Farfor of Lausanne (French), Dr Alan Griffiths and Professor M. M. Willcock of University College, London, and Dr John Richens, formerly of King's College Hospital (Latin and Greek), Mrs J. Haas (Dutch), Dr Elizabeth Cashdan of Pittsburgh, Professor R. D. Martin and Dr S. Jones of University College, London (anthropology), Dr John Nunn of the Clinical Research Centre, Harrow, Middlesex (hieroglyphs), and Professor J. A. N. Corsellis, formerly of the Institute of Psychiatry, London (neuroanatomy). Thanks also to those whose letters of advice and criticism were printed in the *British Medical Journal*, among whom were Professor Sydney Selwyn of Westminster Medical School, Mr R. A. Davis of the Ministry of Agriculture, Fisheries and Food, Professor J. T. Aitken of University College, London, Dr M. F. Hawkins of Dahran, Saudi Arabia, Mr P. J. E. Wilson of Swansea, Dr H. de Glanville of Weybridge, Surrey, and Dr P. Skrabanek of Dublin. I have drawn deeply from the *Oxford English Dictionary*; plunging into its depths greatly increased the enjoyment I obtained in writing this book. Finally, I thank my lifelong companion Digitus Medius Dexter for typing this work.

Note on the second printing: Non-medical readers have shown much interest in the first printing of this book, and I have therefore added a glossary of medical terms used, where these are not already explained in the text. Herein I crave the indulgence of medical colleagues.

Contents

The humours

The belief that good health depended upon the correct balance between various body fluids dates from before Hippocrates' time (*c*.500 BC) and persisted until the middle of the nineteenth century, when it finally succumbed to the cellular pathology of Rudolf Virchow. These body fluids were called HUMOURS (L. *umor*). It is quite proper to apply this term to the fluids within the eye (the aqueous and vitreous humours), and we use the adjective HUMID for anything wet or damp. The revived term HUMORAL refers to the transmission in body fluids of hormones and biochemical mediators.

In ancient times there were thought to be four humours—blood, phlegm, yellow bile, and black bile—and it was the correct balance of these that was supposed to maintain one's physical and mental well-being. If a man was optimistic, self-confident, and sexy, his personality was ascribed to an excess of blood and he was called SANGUINE (L. *sanguis*, *-inis*). If he was cold and unexcitable, he was said to be PHLEGMATIC because of an excess of phlegm (Gk. *phlegma*). If he was easily roused to anger he was said to be CHOLERIC, from an excess of yellow bile (Gk. *cholē*), and, if sad and depressed, MELANCHOLIC because of an excess of black bile (Gk. *melan*, black; *cholē*, bile). This may have been a clinicopathological observation, for depression commonly occurs with obstructive jaundice when the bile in the gall-bladder is concentrated and blackish. Sanguine, melancholic, choleric, and phlegmatic are still used to describe these types of personality.

When we say that someone is in ill or good humour, we maintain the ancient usage of the terms while no longer believing that the mood is due to the balance of body fluids. Current advances in the study of cerebral amines and their manipulation by psychopharmacological agents suggest that the wheel may yet turn full circle.

• Phlegm

Of the four humours believed in former times to be responsible for one's state of health, phlegm is the one most obvious to the layman, at least in its manifestation as mucus in the respiratory tract. Nasal phlegm (mucus) was thought to be a secretion of the PITUITARY gland, supposedly flowing by way of the sphenoidal air sinus into the back of the nasal cavities. *Pituita* is Latin for phlegm; hence, the 'phlegm (producing)' or pituitary gland. The pre-Harveian mind, which could postulate invisible pores in the interventricular septum through which blood flowed from right to left ventricle, would scarcely have regarded the thin plate of bone separating the sella turcica from the sphenoidal sinus as an impenetrable barrier.

The obsolete concept that nasal mucus originates from the region of the brain persists in the French expression *rhume de cerveau* (discharge from the brain) for the common cold (see *Rheum*).

• The -rrhoeas

The so-called humours of earlier times were fluids and by definition, fluids flow. Oddly enough these terms have different derivations despite their similarity; fluid—L. *fluere*, to flow; but flow—via Teutonic from Gk. *ploein*, to float, and

L. *plorare*, to weep. The Greek for flow is *rhein*, whence RHEOLOGY, a branch of applied physics concerned with the study of fluid flow. Rheology has its application in industry, for example in the manufacture of jam and paint, which must not be too runny, nor too thick. Give me thixotropic honey any day!

In the biological sciences, the rheology of cervical mucus is important in relation to fertility, and that of bronchial mucus to expectoration; the non-Newtonian fluid properties of the latter are essential for mucociliary clearance. In medical terminology there are many -RRHOEAS, mostly unappetizing: rhino-, leuco-, gono-, dia-, pyo-, oto-, ameno- and dysmenorrhoea, to name but a few; and they all involve flow. An odd one is CATARRH, derived from a hypothetical catarrhoea, meaning down flow, post-nasal drip and its associations. The -oea seems to have dripped off too.

• Moodiness and the spleen

Black bile was thought to come from the spleen, which thus became known as the organ of melancholy. SPLEEN is still occasionally used to denote an irritable, peevish or morose frame of mind; likewise SPLENETIC to describe this state. The spleen is situated under the costal cartilages (well, it is when enlarged), whence HYPOCHONDRIA (Gk. *hypo*, under; *chondros*, cartilage); hence this region is the seat of melancholy. By a slight shift of meaning we have its modern sense of morbid anxiety about one's health. I suppose no one has THE VAPOURS nowadays, but when they did, this acute emotional disturbance with hysterical manifestations was supposedly due to vapours rising from the spleen and other hypochondrial viscera to affect the brain.

• The liver, kidneys, bowels, and heart

Several viscera were believed to influence emotion, mood and personality. I have previously described the supposed influence of the spleen. In the Middle Ages the LIVER, for no very obvious reason, was thought to be the seat of the passions—that is, emotions in which there is a component of suffering (L. *patior*, pp. *passus*, to suffer). Later it became known as the seat of courage, and a pallid anaemic liver made one cowardly.

> How many cowards . . . who, inward search'd, have livers white as milk.
>
> (*Merchant of Venice*, III. ii. 86)

> For Andrew, if he were opened, and you find so much blood in his liver as will clog the foot of a flea, I'll eat the rest of the anatomy.
>
> (*Twelfth Night*, III. ii. 67)

Perhaps Shakespeare had knowledge of some astute clinicopathological observations here, for Andrew Aguecheek says

> I am a great eater of beef, and I believe that does harm to my wit.
>
> (*Twelfth Night*, I. iii. 84)

Is this hepatic encephalopathy occurring in cirrhosis after a large protein meal?

Though black bile was associated with melancholy, JAUNDICE was associated with envy (green with envy) and jealousy. (Curiously, *jalousie* means venetian blind in French and German. Shades of Maupassant and Schnitzler!). Yellow, too, meant jealous, but towards the end of the nineteenth century it came to mean cowardly.

KIDNEY implied temperament or class ('a man of my kidney', *Merry Wives of Windsor*, III v. 116). BOWELS were the seat of tender and sympathetic emotions ('if any bowels and mercies', *Philippians*, 2: 1). GUTS, implying determination in face of difficulties, is of late nineteenth-century coinage.

Not surprisingly the HEART, as prime seat of the emotions, surpasses all other organs in the richness and variety of phrase and usage to which it has given rise. It owes its prime position to the very obvious sensation of palpitation perceived when one is under acute emotional stress. Love gives us 'dear heart' and 'sweetheart'. Overwhelmed by grief is 'broken-hearted'. Affectionate approval is denoted by 'after one's own heart'. 'Heart of oak' indicates bravery, and encouragement is given with 'Take heart!'. A coward is 'chicken-hearted', implying a small heart. Despondent is 'disheartened', and profound despondency or fear can yield a 'sinking heart'. Sudden fear brings 'one's heart in one's mouth' (? dynamic pulsation of the aorta and its major branches in the neck). Unfeeling and callous is 'heartless', and resistant to the emotional needs of others is 'hard hearted'. To pine is 'to eat one's heart out'. To expose one's feelings to everyone is 'to wear one's heart on one's sleeve', and 'with hand on heart' is a declaration of sincerity. A mutually frank and sincere conversation is 'heart to heart talk'. Occasionally, the heart can think as well as feel, as in 'learning by heart'. Where, I wonder, are the COCKLES of the heart, and why does the application of warmth thereto give satisfaction? Studies in anatomy, physiology and psychology leave these questions unanswered.

Tart, meaning prostitute, was originally 'sweetheart' with no derogatory implications. In the early 1920s it was spelt t'eart, having dropped the 'swee' and 'h'. It meant what would more recently be called bird or chick, and was in a transitional stage as regards spelling and meaning. In such changes of usage lie pitfalls for foreigners; a French lady, wishing to compliment my wife on her cooking, declared 'You are the Queen of Tarts'.

The diaphragm

Close to the heart lies the DIAPHRAGM, and close was its supposed function. Initially, it was regarded by the Greeks as the seat of the soul, and later of the mind—that is, the organ of thinking rather than feeling. They had two words for diaphragm: *diaphragma* and *phren*. *Diaphragma* seems to have referred to its anatomy as a partition separating the contents of thorax and abdomen (Gk. *dia*, across; *phragma*, partition), while *phren* seems to have referred to its function of thinking. We still use *phren* in anatomical terminology, such as costophrenic and cardiophrenic angles, and phrenic nerve. The concept of phrenic pertaining to the mind persists today in frenetic, frantic and frenzy—though all are to do with wild excitement rather than sober thought. The root was revived by Franz Josef Gall in the late eighteenth century when he coined the term 'phrenology', a pseudo-science that was practised until the early years of the twentieth century.

Fascia

FASCIA is a sheet of fibrous tissue which envelops the whole body under the skin, and also individual muscle groups. It is so named because of the band-like arrangement of its constituent fibres. Fascia means band or bandage in Latin, and modern Italian; it is curious that the Italian pronunciation of -sci- as 'sh' is used. A FASCICLE is a bundle of sheets of paper comprising a part or instalment of a printed work. A fascicle, anatomically speaking (and then usually in Latin usage as FASCICULUS, diminutive of L. *fascis*, bundle) describes a bundle of muscle or nerve fibres, many of the latter being specifically named. FASCICULATION is repetitive twitching of a muscle fasciculus. The flat band-like instrument board of your car, the FACIA, is a distant collateral.

While we are on the subject, FASCIST has the same root. In 1915 Mussolini founded the *Fasci d'Azione Rivoluzionaria* (Band of Revolutionary Action) and, when he took power in 1922, he adopted the fasces as the emblem of his party. The fasces comprised a bundle of sticks with a projecting axe-head, which was carried in the days of ancient Rome by attendants (lictors) of senior magistrates as symbols of authority (Fig. 1). Mussolini believed he would be instrumental in reviving the glories of ancient Rome.

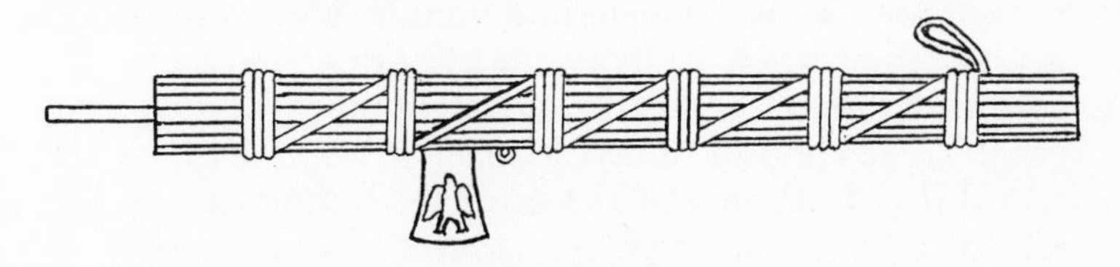

Fig. 1 Fasces, a bundle of sticks with a projecting axe-head: the symbol of authority in ancient Rome.

• Nerves and tendons

No distinction was originally made between NERVES and TENDONS—not surprisingly because both are predominantly white, thonglike structures. Both were called neuron, whence we have neuron(e), neurology, neuropathy, neurosis, and so forth. One type of neurosis, however, is tendinous, namely an APONEUROSIS. This, being a flat sheet, is oddly the opposite of thonglike. The expression 'to strain every nerve', meaning to use one's utmost endeavours, refers to the nerves as sinews, whereas when 'nerve' denotes courage and, in recent years, an element of cheek ('You've got a nerve!'), the neurological variety is implied.

• Arteries and airways

ARTERIES were so called because they were believed to contain air. They were seen to be empty after death, so they were obviously not blood conduits. They were believed to convey 'vital spirit' (Gk. *pneuma*), a material derived from inspired air and not clearly distinguished from it. The usage persists also by implication in the word TRACHEA, Greek for the rough; the term is short for *arteria trachea*, the rough air-carrier, rough because of the cartilaginous rings. The noun was later dropped, the adjective persisted and became a noun. This is a commonly occurring process in word evolution; after all, we have 'the rough' in golf. It really was all quite simple. Air was taken in by the *arteria trachea*, endowed with vital spirits on mixing with blood in the heart whence it pulsed to and fro in the arteries (pneumatic arteries?).

• Futurology today

Mankind has always attempted to foretell the future. The Romans had their own complex systems of divination. I suspect that the recent revival of interest in astrology is symptomatic of the decline in religious belief during the past half-century; astrology now provides for its believers the feeling of security that was thereby lost. Be that as it may, scientifically educated people and those able to face the uncertainties of the future reject the dogmas of astrology as they do divination by cards, palmistry, ouija boards, crystal balls, and tea-leaves. Yet we still use words derived from ancient beliefs in omens and in prognostication therefrom. We have all heard something like this.

> On this auspicious occasion we are honoured to have with us the eminent speaker who will inaugurate the new academic year. We considered him the obvious choice, and felt it would have been nothing less than a disaster if he had been obliged to decline our invitation following a severe bout of influenza.

If this sentence sounds a bit contrived, that's because it is. Let us look at auspicious, inaugurate, consider, disaster and influenza. AUSPICIOUS is from L. *auspex*, a contraction of *avispex*, which is derived from *avis*, bird; *specio*, to look; divination by observing the flight of birds. INAUGURATE, originally to usher in with auguries. In ancient Rome the augur was an important religious official who interpreted omens derived from the flight, feeding, or singing of birds; the appearance of the viscera of sacrificial victims, and other portents; probably from L. *av(is)*, bird; *gar(rio)*, to chatter. In Italian, *Auguri*! (good wishes for the future, Good Luck!) retains some of its original meaning. CONSIDER comes from L. *con*, and *sidus*, *sideris*, star; a star or constellation as a guide to a course of action. DISASTER an unfavourable aspect of a star or planet, from L. *dis*, privative; *astrum*, star; that is, ILL-STARRED. INFLUENZA is from the Italian *influenza*, influence; an epidemic disease due to astral influence. I always used tea-leaves, but have learnt to face the future with resolution since we changed to tea-bags.

• Drug

The word DRUG is undergoing a change of meaning. Until the middle 1960s 'drug' meant a pharmacologically active medicinal substance. Only in special context did it denote a substance which affects mood and thinking, with the attendant risk of addiction. 'Drug' is now generally understood by the laity in the latter sense and is, alas, fast becoming a dirty word. Even the Committee on Safety of Drugs has been renamed the Committee on Safety of Medicines. It was an eminently usable word in clinical practice; monosyllabic, four-lettered, clear, and to the point. Nowadays one hesitates to ask a patient 'What drugs are you taking?' for fear of being misunderstood. 'Medicines' can be understood, especially by older patients, to mean only fluids; for example, two teaspoonsful after meals. So, instead, one now takes a deep breath and says, 'What medicines, tablets, or capsules are you taking, or—er—sprays— ahem—suppositories or, excuse me—pessaries?' 'None of these, doctor', (I knew I'd forgotten something.) 'I'm on injections'.

The present trend is a reversion to earlier times when the use of 'drug' could optionally imply narcotic effect. 'I have drugg'd their possets' (*Macbeth*; *II*. ii. 7).

Drugs and their names

It is a pity that medical dictionaries omit the derivations of drug names, because the exploration of this field is quite fascinating, although too large for a detailed account in these short notes. Fortunately, the pharmaceutical names of most drugs are based on terms in organic chemistry. Here are the derivations of a few common building-blocks from which terms in organic chemistry are constructed.

- OL suffix denoting an alcohol. 'Alcohol' is derived from the Arabic *'al kohl'*, meaning cosmetic eye-shadow made from finely powdered antimony sulphide obtained by sublimation. Later, it was applied to any substance obtained by sublimation or distillation including 'alcohol of wine'. Eventually restricted to this substance, the words 'of wine' were dropped.

METHYL Gk. *methy*, wine; *hylē*, wood; wood spirit.

ETHYL Gk. *aither*, the clean upper air. Originally adopted for diethyl ether (anaesthetic ether) because it was volatile and clean-smelling.
PROPYL Gk. *pro*, before; *pion*, fat. So called because it is the first of a series of acids whose derivatives are oily (the fatty acids).
BUTYL Gk. *boutyron*, butter. Butyric acid is formed in rancid butter.
AMYL L. *amylum*, starch. Its alcohol was first obtained from *fusel* (Ger. rotten, bad) oil separated in rectifying (purifying) alcoholic spirits distilled from fermented grain or potatoes.
ACET- L. *acetum*, vinegar which contains about 6 per cent of acetic acid. Note in passing that ACETABULUM is a little vinegar cup.
CITR- L. *citrus*, lemon, the juice of which contains about 6 per cent of citric acid.
FORM- L. *formica*, ant. Formic acid is present in ant stings. It was so named by John Ray (1670), who obtained it by distillation of red ants.
LACT- L. *lac*, *lactis*, milk. Lactic acid is formed in sour milk.
OXAL- oxalic acid was so called because it was first prepared from oxalis, the wood sorrel. This in turn was so named because of the sharp, acidic taste (Gk. *oxys*, sharp, acid) of the leaves.
CYAN- Gk. *kyanos*, dark blue, from the blue colour of ferric ferrocyanide, also known as Prussian blue after its discoverer Diesbach, an eighteenth-century artist and colour-maker in Berlin. Incidentally, also, the blue of the original blue prints for making copies of plans, an expression now only used metaphorically.
ALDEHYDE al(cohol) + dehyd(rogenatum). Because it is made by oxidation of an alcohol.
PHEN Gk. *phaino*, to appear, shine. Phenolic and other coal-tar substances were by-products of the manufacture of gas from coal for illuminating purposes.
BENZ Arab. *luban-jawi*, frankincense of Java, the fragrant resin obtained from a Javanese tree. *Luban-jawi* gave *lo-benjuy*,

from which the *lo* was dropped in the mistaken belief that it was a grammatical article; thereafter from benjoin to benzoin, which contains benzoic acid esters.

ESTER contracted by Gmelin from Ger. *Es(sigä)ther*, acetic ether, to distinguish it from ordinary ethers, since it was a compound of an alcohol and an acid radical.

AZO denotes the presence of nitrogen (F. *azote*) in a compound. *Azote* was so called because it did not support life (Gk. *a-*, not; *zōos*, life).

THIO Gk. *theion*, sulphur.

AM (-INE, -IDE) ammonia was the name given by T. Bergman in 1782 to the gas obtained from sal ammoniac. Sal ammoniac, or salt of Ammon, was obtained from the dung of camels near the temple of Jupiter Ammon (the Egyptian god Amun) in Libya.

Drugs from the New World

A few drug names come from the Indians of the Americas. Tupí-Guaraní is still spoken by the indigenous people of Brazil. In this language *ipe-kaa-guene* means 'the low plant causing vomiting'. An extract from its root was called IPECACUANHA by the Portuguese. It is used as an emetic and expectorant.

JALAP derives from Sp. *purga de Jalapa*, a purgative obtained from the roots of an ipomoea plant growing in the vicinity of Jalapa, a city of Mexico. The Aztecs called their city *Shalapan* which, in their language, Nahuatl, means 'water upon sand'.

Kiché, the language of the Mayans of Guatemala, gives us QUININE. Their *kina* (spelt *quina* in Spanish) means bark. The duplication of a plant name almost invariably implied medicinal properties, hence *quinaquina*; the suffix '-ine' is commonly adopted for derived substances. This is not to be confused with the CINCHONA tree from which quinine was obtained. It was so named by Linnaeus in 1742 in honour of the Duchess of Chinchón (a small town 40 km from Madrid). In 1638, when vice-queen of Peru, she was cured of malaria by the use of the bark, which she brought to Europe two years later.

CURARE, the arrow poison that kills by paralysis of breathing (p. 14), derives from the Macusi Indians of Guyana. Pedants for precision in pronunciation will enjoy knowing that the initial c represents a click. Curare has nothing to do with the identical Latin word meaning to care for—as a curate cares for the souls in his parish. As people used to say in bygone days, when praising the family doctor, 'He cured my father till he died'.

The drug shop

APOTHECARY is the earliest English term for those who prepare and sell drugs. It is now archaic except for its use by the Society of Apothecaries, a body of distinguished doctors who confer a licence to practise medicine, the LMSSA. The term is derived from Gk. *apo*, away; *thesis*, put; *apothēkē*, a place where things are put away, a store-house. 'Apothec' is obsolete in English but foreign equivalents are currently used; Ger. *Apotheke*; Scand. *apotek*; Dutch, *apotheek*. The French *boutique*, a special kind of shop, and the Spanish *bodega* a store-room (hence a wine cellar) also stem from *apothēkē*.

CHEMIST, short for pharmaceutical chemist, was the term used until recently, and it is still so used by the general public in the United Kingdom. Chemist derives from alchemist, alchemy in the Middle Ages being the pursuit of the transmutation of base metals into gold. This in turn derived from the Arabic *al khimia*, of obscure meaning. According to Mr A. L. Pahor, the derivation is from *Kimi* (Egypt, in ancient Egyptian).[1] PHARMACIST, a word of Greek derivation (*pharmakon*, a drug or poison), has been the preferred professional term since the 1940s, and similar terms are used in all Romance languages (F. *pharmacien*). The Spaniards use both forms, *farmacéutico* and *boticario*, who work respectively in a *farmacia* and a *botica* (a *bodega* for drugs, as it were). DRUGGIST is of mainly Scottish and American usage. The word PHARMACY has not caught on with the British lay public. They still like to get their drugs—sorry!, medicines—at the DISPENSARY in the hospital and at the CHEMIST'S in the high street.

Bows and arrows

The word PHARMACY derives from the Greek *pharmakon*, a drug or poison. Many drugs must have been poisonous in ancient times, as now. Interestingly, the Greeks had one word for both. The safest epoch pharmaceutically seems to have been the century 1840 to 1940 when, with few exceptions, drugs were neither poisonous nor effective.

Special interest attaches to the derivation of the words TOXIC and TOXIN. In Greek, *toxon* is a bow (the sort that flings an arrow) and *toxikos* is its adjective. Toxophily is the sport of archery. The Greeks must have known of arrow poisons, but they called them bow poisons, *toxikon pharmakon*. As so often happens, the noun was dropped and the qualifying adjective remained to become a noun in its turn; thus toxicos, of a bow, came to mean poison; hence toxic and toxin.

Three arrow poisons have come into use as drugs, and add an extra etymological flavour to the use of the root *pharmakon*. CURARE, still used by the natives of the South American jungle, was introduced as the first muscle relaxant for anaesthesia. Nowadays synthetically produced curare-like compounds are used for this purpose. STROPHANTHUS and OUABAÏNE are cardiac glycosides of the digitalis group, obtained from the seeds of African plants.

[1] Pahor, A. L. (1981). More about ancient Egyptian (correspondence). *Br. Med. J.* **282**, 484.

• Pills and tablets

Pills and tablets have two things in common—they contain a single dose of drug and are designed to be swallowed whole. In other respects they differ. A PILL (L. *pilula*, diminutive of

L. *pila*, ball) is a uniformly medicated sphere which is made by mixing a drug with inert material (the excipient) to form a mass that is plastic (in the sense of having a doughy or putty-like consistency). It is then rolled into shape and finally coated with a varnish to protect the surface. The procedure is time-consuming in relation to the quantity of final product, and does not lend itself to mass production. For this reason the pill has been superseded by the tablet and the capsule in medical practice, though it is still made for herbalists and homeopaths. By contrast, the TABLET (diminutive of table; L. *tabula*, board or plank) is in general terms 'a small flat and comparatively thin piece of . . . hard material artificially shaped for some purpose; a small slab' (*OED*); hence, in a pharmaceutical sense, it may be described as a small compressed mass of medicament, usually circular, and flat or biconvex. While pills are of great antiquity, tablets are of comparatively recent origin. Although a compressed tablet was patented in the middle of the nineteenth century (Brockendon, 1843), they were not made in any quantity until early in the twentieth century. I venture to suggest that not one doctor in a hundred who qualified after 1945 has seen a pill.

The expression 'pill-rolling' has been used to describe the thumb and finger movements of Parkinsonian tremor. This is curious because, since the first quarter of the nineteenth century, the great majority of pills were rolled between two flat wooden surfaces especially designed for the purpose. James Parkinson in his 'Essay on the Shaking Palsy' (1817) makes no mention of pill-rolling; nor does Gowers in 1888, or James Collier and W. J. Adie in the 1922 first and later editions of Price's *Textbook of the Practice of Medicine*. Gowers came fairly close in describing 'the thumb and forefinger, which may move as in the act of rolling a small object between their tips'. Interestingly, Price described a 'drum tapping' movement which well illustrates the alternating pronation-supination. 'Pill-rolling' appears in Kinnear-Wilson's monumental work (1940) but within quote-marks as though he did not quite approve, and with the comment that the term was 'now in vogue'. This suggests the phrase gained ground in the 1930s, when perhaps the last few pills were rolled by hand.

And why have I differentiated, it may seem somewhat pedantically, between the pill and the tablet? It is the current misuse of the former term in place of the latter, at first by the laity and now even by doctors, especially those who have never seen a pill! It all began with 'THE PILL' which is currently the most widely used term for an oral contraceptive. 'The Pill' is in fact not a pill but a tablet. However, 'The Pill' has caught on and 'to be on the pill' can mean only one thing.* It must be admitted that, with present usage, 'Are you on the tablet?' would sound a bit awkward.

• Fatal or lethal?

Drugs, poisons, diseases, and injury which cause death are said to be fatal or lethal. Why fatal? In Greek mythology there were three Fates; these ladies controlled the thread of life. Clōthō spun the thread and Lachesis mixed the strands of good and evil fortune. She passed it on to Atropos who cut the thread of life. *A-tropos*, no turning; she who would not be turned aside from her task. Hence, atropine from *atropa belladonna*, the deadly nightshade. The Fates were concerned not only with death, but with birth and the course of one's life. So perhaps

* I believed this to be true when the article appeared in the *British Medical Journal* in January, 1981. Unfortunately, to some people it may mean any pill or tablet. W. Macredie, writing in the same journal in 1983, described a young woman who admitted she was on the pill. After surgery for acute appendicitis she became comatose. The pills, which had been withheld, were chlorpropamide and the coma was diabetic.

fatal is not a suitable word for causing death, and lethal (L. *letalis*; *letus*, death) is the better choice. There is a probable connection with Gk. *lēthē*, forgetfulness; whence the river Lethe in Greek mythology, a river of the infernal regions from which the souls of the dead drank, whereupon they forgot their previous life.

• Guano

GUANO is the accumulated excreta of sea-birds, and by extension, of any birds and of bats and seals. Massive deposits (over two hundred feet thick on some islands off South America) are worked, to be used as fertilizer, as guano is a useful source of nitrogen, phosphate, and potassium. Guano is a Spanish word taken from *huano*, dung, in the Quechua language. At the time of the Spanish conquest of the Incas, Quechuan was spoken along the east coast of South America from Colombia, Ecuador, and Peru as far as mid-Chile, and it is still spoken by several million people. In 1850 a compound was extracted from guano, bearing some resemblance to the xanthines. Accordingly, its discoverer[1] named it GUANINE. A century later Watson and Crick discovered that guanine is one of the four basic substances that are integral in the molecular structure of DNA.

GUANIDINE, originally prepared from guanine, has several derivates of industrial use. It is also a molecular component of creatine and arginine, and of the antimalarial drug PROGUANIL, the hypotensive drug GUANETHIDINE, and the intestinal antimicrobial SULPHAGUANIDINE.

[1] Fownes, G. (1850). *A manual of elementary chemistry* (3rd edn). John Churchill, London.

• Balsam

A BALSAM is a viscous, oily, or resinous substance of vegetable origin. The Greek *balsamon* appears to derive from the Hebrew *besem*, balsam-odour or, less probably if more picturesquely, *ba'al shemen*, master of oils. Balsam of Peru was used as an antiseptic dressing, and Balsam of Tolu (a town and district of Colombia, just over the border from Peru) was taken orally as an expectorant. The medicinal use of these balsams is obsolete in western medicine, but Friars' Balsam is still used as a hot inhalation for respiratory infections. From balsam we have the word BALM, a fragrant oil or ointment with soothing properties, and today used more in a metaphorical than literal sense; hence balmy (not to be confused by the southern English with barmy). To EMBALM is to preserve a cadaver from putrefaction. Though nowadays this is done by perfusing the blood vessels with formalin, the ancient Egyptians (Fig. 2) used balsams, most of which contain benzoic acid (see *Benz*, under *Drugs and their names*). This is probably the main ingredient in achieving preservation of the tissues; benzoic acid has antimicrobial and antifungal actions and is now widely used as a preservative in the food and soft drinks industries, in sodium benzoate (E211) and various other benzoates.

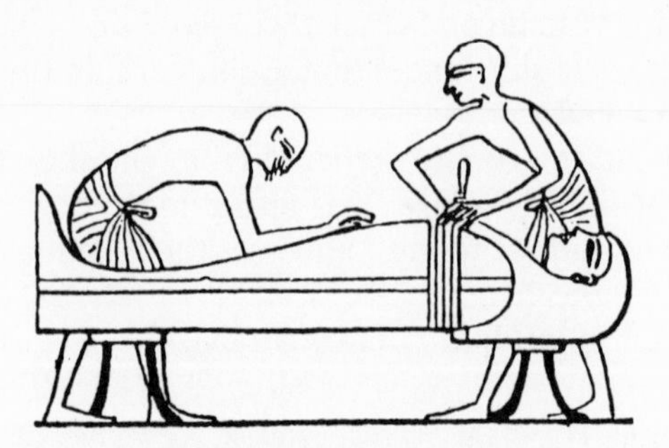

Fig. 2 Egyptian embalmers at work.

The place name Tolu in Colombia crops up in TOLUENE, which was first obtained by distillation of Balsam of Tolu. We meet toluene in its presence as part of larger molecules. Butylated hydroxytoluene (E321) is an antioxidant and is used to prevent the development of rancidity in stored oily foods. Repeated exposure to toluene diisocyanate during the manufacture of plastic foams has induced asthma in a proportion of factory workers.

• Flu-, fluo-, fluor-, fluorine, fluorescence, flux

These terms all originate from fluorspar. Spar is a general term for several bright, crystalline minerals admitting of easy cleavage (*OED*). Fluorspar (calcium fluoride) was prefixed 'fluor' because of its fusibility (MP 1330 °C) and its use as a flux in soldering and brazing (L. *fluo*, p.p. *fluxus* to flow; see *The -rrhoeas*). FLUORINE was so named by Ampère (1810) in anticipation of its isolation from fluorspar; it is the most highly reactive of the halogens. The inclusion of its atom in the molecules of many synthetic organic substances that are used as drugs (notably steroid hormones and diuretics) increases the pharmacological activity. The prefixes 'fluor-', 'fluo-', and 'flu-' are encountered with increasing frequency in drug names.

Fluorspar emits a blue or green light when illuminated by ultraviolet light; this optical phenomenon (which occurs with many other substances) is known as FLUORESCENCE. Finally, we have the outmoded medical terms FLUOR ALBUS, later Graecized to leucorrhoea, for white (vaginal) discharge, and FLUX, meaning a fluid discharge from a wound (serum, blood or pus) or from the bowel.

• Calcium

The Latin *calx*, *calcis*, has some interesting medical derivations; it means pebble, stone, or limestone, whence we have CALCIUM, CALCAREOUS, CALCIFICATION, and CHALK. CALCULUS, a little stone, is the diminutive of *calx* and is applied to concretions occurring in the body—for instance, gall-bladder and urinary tract. Little stones were used by the Romans for reckoning on the counting board and abacus, whence we have calculus in mathematics and CALCULATE, to compute mathematically; to opine (American usage). Incidentally, concretion has nothing to do with L. *creta*, chalk; it derives from L. *con*; *cresco*, p.p. *cretum*, grow together. Concrete also stems from this root.

Calculus in dentistry is a calcareous concretion precipitated from saliva on the surface of teeth. An older term for dental calculus, still used, is TARTAR. The word tartar was originally applied to the hard crust forming on the sides of wine casks during fermentation, whence by analogy it was adopted by dentists (1806). On analysis the deposit on wine casks proved to be the potassium hydrogen salt of an organic acid, which was consequently named TARTARIC. The same salt, known as CREAM OF TARTAR, was formerly used as a laxative.

Calx, *calcis*, meaning heel in Latin, (which is of entirely different origin from *calx*, pebble) gives us OS CALCIS and CALCANEUM, the heel bone, and L. *calcar*, spur, whence CALCARINE cortex and fissure, from a fancied resemblance of that cerebral cleft to the aforesaid item of equestrian equipment. In Roman times a soldier's shoe was *caliga*, from *calx*, heel. The diminutive of *caliga* was *caligula*, a little shoe of this type, perhaps for children dressing up. Thus, Caligula became the nickname in his boyhood of Gaius Caesar, who, in current slang, might well be termed a 'heel' considering his later behaviour as emperor.

• Saxifrage

Surgical removal of bladder and kidney stones has been the method of choice because the underlying cause may possibly be eradicated at the same time. (Many patients who would formerly have undergone surgery are now treatable with extracorporeal shock lithotripsy.) However, no patient likes operations, and in bygone days removal of bladder calculi by the perineal route ('cutting for stone') must have been anticipated with horror. Not surprisingly medical (in the sense of non-invasive) remedies were sought. Herbs of the genus *Saxifraga* were formerly used for the treatment of stone and gravel, in the belief that they would break the stone into fragments, hence the name (L. *saxum*, rock; *frag-*, root of *frangere*, to break). The notion that the plant had this property probably stemmed in the first place from the tendency which some species exhibit to grow among stones and in clefts between rocks; indeed, they are now cultivated in rock gardens. An alternative suggestion about the basis for the esteem in which saxifrage was held refers to the tubercles on the root of *Saxifraga granulata* (white meadow saxifrage) which were likened to fragments of stone.

The erstwhile doctrine of signatures was based on the hypothesis of curing like with like. This latter suggestion was a later belief. However, as long ago as the first century, Pliny the Elder wrote about saxifrage in his *Natural History*, 'It breaks stones and dislodges them from the body wonderfully'. Wine-lovers suffering from renal calculi will unfortunately derive no special benefit from drinking Niersteiner. This wine comes from vineyards in the neighbourhood of Nierstein, a town on the left bank of the Rhine a few kilometres south of Mainz. Nothing to do with *Nierensteine* (Ger. kidney stones). Just enjoy the wine!

Jade

JADE is a tough, fine-grained semiprecious stone; it is so hard that it cannot be scratched by steel. It has been highly prized by the Chinese since the earliest days of their civilization, and widely used by lapidaries for the cutting of gems, and for exquisite carvings. The first carved jades to reach Europe were brought back by early Spanish navigators from their colonies in Mexico and Central America, where they were used for ornamentation. Among the Spaniards jade pieces acquired a reputation for the prevention and cure of kidney diseases, possibly because some of the native carvings retained the reniform character of the original jade pebbles. Hence they were called *piedra de los riñones* (stone of the kidney) or *de ijada* (of the flank or loin), and it is the latter that gives us 'jade'. In the eighteenth century nephrite was coined as an alternative name, possibly to emphasize the renal connection in a more terminologically recognizable form. In 1863, A. Damour showed that specimens of 'jade' were one or other of two unrelated minerals, though of similar appearance and physical properties; he accordingly named them 'jadeite' (sodium aluminium silicate) and 'nephrite' (calcium magnesium silicate)[1]

[1] Hansford, S. H. (1963) Jade and other hard stone carvings. *Encyclopaedia Britannica*. William Benton, London.

The drink problem*

Q Have you a drink problem?
A I have no difficulty in swallowing liquids.

* With apologies to Jaroslav Hašek, author of *The Good Soldier Švejk, and his fortunes in the world war*. (1922) William Heinemann, London. Translated from the Czech by Cecil Parrot, 1973.

Q I mean, have you a problem with alcohol?
A No, I can get all the booze I want.
Q I mean, do you drink more alcohol than you should?
A Why? How much should I drink?
Q What I mean is, do you drink more alcohol than is good for you?
A That depends on how much is good for me. What quantity do you recommend?
Q (sigh) Is alcohol affecting your behaviour for the worse?
A My friends don't think so, but my wife is inclined to fuss about it.
Q So your wife has a problem with *your* drinking.
A No problem at all. She is quite clear in her mind what is wrong.
Q Thank you! I will see you again next Thursday. (Slips off his elastic-sided shoes which have begun to feel uncomfortably tight.)

• A gem-stone for sobriety

It has long been recognized that the immoderate enjoyment of alcohol is counterbalanced by disagreeable sequelae. Various methods of enjoying alcohol without getting drunk have been tried, the intention being to yield a modest plateau rather than a peak effect. Prior ingestion of milk or other fatty substances delays the exit of alcohol from the stomach; fructose increases its rate of catabolism. Caffeine, traditionally taken as black coffee for rapid absorption, is a cerebral stimulant and may yield a hoped-for pharmacological antagonism.

The problem is not new. To the ancient Greeks, Bacchus embodied the sociable, jolly, even wildly hilarious, frame of mind induced by wine-drinking, while his tutor, the fat, drunken Silenus, seems to have shown the unwanted effects. The Greeks seem to have explored every aspect of human behaviour, and had a simple way of dealing with the problem, provided one had the equipment. Drunkenness could be prevented by drinking from a cup made of purple quartz. They called it AMETHYST, from *a-*, not, and *methystos*, inebriated (*methy*, wine).

• Clinic

CLINIC has undergone a reversal of meaning. In its modern sense it denotes an out-patients' department, a place in or attached to a hospital where a doctor sees patients who are not resident at the hospital and who are definitely not in bed. But Gk. *klinē* is a bed, a place where one re/clines. 'Clinic' was first used in relation to a sick-bed in the seventeenth century. A clinical lecture (1720) was one given at the bedside. In the nineteenth century a clinic was a private hospital. Then in 1892 the term was first used to describe an institution attached to a hospital at which free treatment was given, whence the present-day meaning was derived. Etymologically it would be more appropriate if the word clinic were applied to the ward-round, because there one will find the patient (with any luck) in bed.

CLINICAL has been adopted by the laity to describe an unemotional, scientific, technically efficient approach to a task; one uncontaminated by consideration for the feelings of those concerned. This usage is probably of journalistic provenance. For an entertaining and witty account of the variations on this theme one cannot do better than read the 'Personal View' by

Professor D. H. Smyth in the *British Medical Journal* of 25 January 1975, p. 203.

• Ward round

Why 'round'? Almost all wards are rectangular, and one's progress during a ward round is along one side, across the width at the far end, and back up the other side, which is anything but circular. I am reminded of a nursery rhyme in which a poor fellow who had lost his way, was advised to go 'straight down the crooked lane and all around the square'. In recent years the course of a ward round has deviated even more from the circular. Such is the difficulty at times of finding empty beds for emergency cases that a round may involve visiting a series of patients in different wards in widely separated parts of the hospital, and the course of a round is that of a three-dimensional zig-zag over the best part of half a mile (0.8 km, no less). Then there is the Grand Round where no one goes round, and everyone is seated except the doctor expounding in what is, in effect, a case demonstration. There are few who have been privileged, like myself, to do truly circular ward rounds. This was some years ago at St Giles' Hospital in Camberwell, south-east London, where there is a drum-shaped block of circular plan. On each floor about 20 beds were arranged radially like those in a bell tent. Ward rounds here were both clinical and topographical.

There is a current trend to rename the round 'ward conference', perhaps on the grounds of being a more appropriate functional description, and because of the sense of importance conferred by Latinization and the two extra syllables. There seems to be no good reason to discard 'round' since everyone knows what it means.

• Stick to Greek and Latin roots

We are lucky that English is an international language in technical and scientific affairs. It lessens the need to learn foreign languages, while imposing a corresponding burden on those whose mother-tongue is not English. The established practice of using classical Greek and Latin roots in the formation of new medical terms gives them widespread intelligibility, as speakers of Romance languages will understand all the Latin-based and some of the Greek-based words, while many of the university-educated speakers of Teutonic languages will have some knowledge of these ancient languages. In coining new medical terms we should, within reason, stick to this custom.

When I see words such as 'breakdown' for analysis, 'see-through' for transparent, 'set-up' for organization, 'blow-up' for (photographic) enlargement, and 'hold-up' for obstruction— all acceptable in brisk, graphic, everyday speech—I begin to wonder whether this trend could be the thin end of the wedge to penetrate medical terminology. But does it matter? See what has happened to the German language and let that be a warning. New words formed from Latin or Greek have often been translated root for root into Teutonic equivalents, thus erecting an unnecessary barrier to international comprehension. A few examples follow, with the centre column showing a literal translation.

hydrogen	water-stuff	Wasserstoff
carbon	coal-stuff	Kohlenstoff
television	far-seer	Fernsehen
technology	industry-science	Gewerbekunde
aluminium acetate	vinegar-acid clay-earth	Essigsäure Tonerde
anaemic	blood-poor	blutarm
appendicitis	blind-bowel-inflammation	Blinddarm-entzündung
duodenum	twelve-finger-bowel	Zwölffingerdarm

I have heard that mitral stenosis became *Mitralstenose* only by the skin of a cusp, as it were, having escaped from a recommendation that it be named *Bischofsmützeklappenunzugänglichkeit*, bishop's-cap (mitre)-valve-inaccessibility.[1]

[1] Smyth, D. H. (1973). Medically oriented language courses. *Br. Med. J.* **4**, 236.

• Ambi- and amphi-

Ambi- and *amphi-* are Latin and Greek respectively for around and about, on both sides. L. *ambo* and Gk. *amphō* mean 'both'. *Ambi-* gives us:

AMBIENT, surrounding (air, temperature).
AMBIVALENCE (L. *valens*, having strength or worth), the co-existence of antithetical attitudes or emotions.
AMBIDEXTROUS (L. *dexter*, right), having the skill of a right hand on both sides.
AMBITIOUS (L. *eo*, p.p. *itum*, to go), to go around, to canvas for votes; hence having the desire for self-advancement.
AMPUTATE (L. *amputo*, p.p. *amputatum*; from *am(b)*, about, with *putare*, prune, lop, as the branches of a tree in pruning).
REP. AMBO is used for 'Repeat both' by those who still prescribe drugs in abbreviated Latin.

Amphi- gives us:

AMPHITHEATRE, an oval or circular auditorium on both sides of the central arena and, in ancient Rome, the Imperial podium (Fig. 3). This is in contradistinction to the Roman THEATRE which had a semicircular auditorium on one side of a diametrically sited stage. The term amphitheatre is sometimes misapplied to a semicircular auditorium with steeply raked seating which exists in some older lecture theatres. It is now

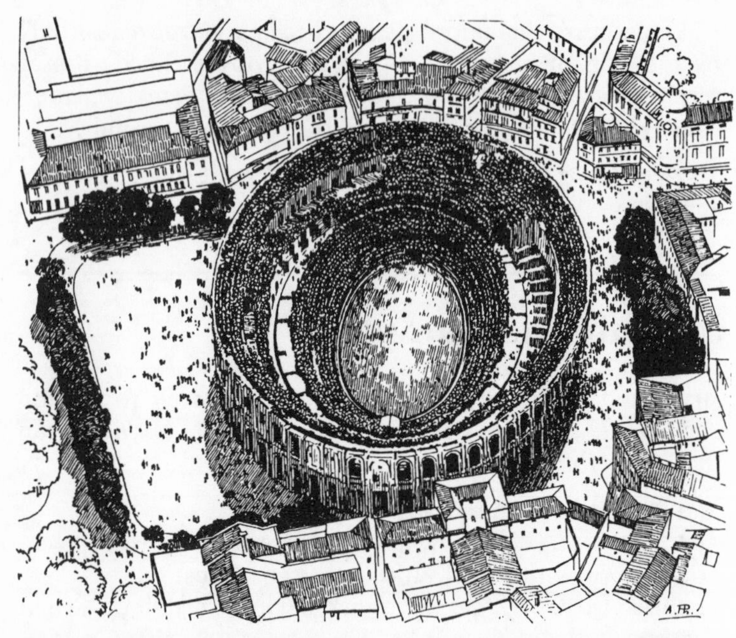

Fig. 3 The Roman amphitheatre at Nîmes, Provence, France. (Courtesy of Michelin, from their Tourist Guide 'Provence', 3rd edn, 1969.)

an accepted, though topographically incorrect, term for one of the galleries in some modern theatres used for entertainment. AMPHOTERIC (Gk. *amphoteros*, on both sides), having both of opposite characteristics, such as reacting as either acid or base. AMPHIBIOUS (Gk. *bios*, life), living both on land and in water. AMPHORA (Gk. *phoreus*, carrier), a flask or jar with handles on opposite sides. Amphora give L. *ampulla*, whence probably AMPOULE. Amphoric breathing denotes breath sounds (on auscultation of the chest) whose quality resembles that of the sound emitted by blowing across the mouth of a jar, usually indicating a large cavity. Two handles are not necessary for this manoeuvre, but they make it easier to hold the jar (Fig. 4).

Fig. 4 Amphora—Greek form, sixth century BC. (From *The Macmillan Encyclopedia*, first published 1981 by Macmillan, London. Courtesy of Market House Books Ltd., Aylesbury, UK.)

AMPHETAMINE is not a derivative, however. This is an acronym derived from the drug's chemical name—*a*lpha *m*ethyl *ph*ene*t*hyl *amine*.

AMBULANCE is a special case. To ambulate, to amble, is to walk around and about. *L'hôpital ambulant* (walking hospital) in early or mid-eighteenth century France comprised the military surgeons and their assistants who followed the army, and gave treatment on the spot. *L'hôpital ambulant* was replaced by F. *ambulance* (1762) and the term then referred to a mobile vehicle with the accompanying staff. It was Napoleon's great military surgeon, D. J. Larrey, who revolutionized the management of the wounded by rapidly transferring them from the fighting zone to base, using his *ambulance volante* (flying ambulance; Fig. 5).[1] 'Ambulance' was adopted in English about this time, but did not come into general use until the Crimean war. As often happens with the drift of language, the noun is dropped and its qualifying adjective or present

Fig. 5 Ambulance, early nineteenth century: Larrey's *ambulance volante*. (Courtesy of *Proceedings of the Royal Society of Medicine*, London.)

participle becomes the noun (*ambulant* to *ambulance*). As ambulant in a medical context means able to walk, we have the seeming paradox that an ambulance is for the conveyance of patients who are not ambulant. A similar relationship exists for the pram, contracted from PERAMBULATOR, but here the nursemaid perambulated the non-ambulant infant.

[1] Richardson, R. G. (1977). Larrey—What manner of man? *Proc. Roy. Soc. Med.* **73**, 490.

• Neologopoiesis

New medical and scientific words are constantly being coined, and are usually constructed from Greek or Latin roots. The classical origin has the force of tradition and in some measure ensures international acceptability. Richard Asher was concerned that new terms should be properly constructed so, with tongue in cheek, he himself coined the word 'neologopoiesis' for new word formation.

Anyone feeling the urge to indulge in neologopoiesis should consult an expert in the ancient languages, unless he or she is that rare person, a doctor or scientist with a good knowledge of Latin and Greek. Try approaching the academic staff in that department of your alma mater; failing that, the classics teacher in the local high school. Doing it from a dictionary has its pitfalls, and may yield something akin to the puzzling instructions that went with some Japanese electronic equipment in the 1960s.

Generally, the adoption of classical roots is straightforward. The word GAS is unusual, not only because of the author's groping towards an understanding of the concept, but also because of the phonetic shift that occurred later. The Belgian chemist and physician van Helmont (1577–1644) made an identification of carbon dioxide, and was the first to understand that there are gases other than air. Needing a word for this, he took the Greek *khaos*, meaning infinite space or chasm (whence also the word CHAOS). The initial letter of *khaos* is the Greek letter chi, pronounced as the ch in loch. Since he wrote in Latin, but within a Dutch-speaking environment, this was further transliterated to 'gas', as the Dutch g has the phonetic value of the Scottish ch. But in all other languages g followed by a is 'hard', as in gap.

Isotope and atopy are words that were less well designed than they might have been. ISOTOPE (1912), from Gk. *isos*, equal; *topos*, place, would have been better as homotope (Gk. *homos*, same; with *topos*) because it occupies the same place in the Periodic Table of the elements. ATOPY (Coca, 1923) (Gk. *atopos*, out of place, strange, odd; from *a*-, no; with *topos*) was applied to a group of allergic diseases, for instance eczema, asthma, and hay-fever, some of which commonly occur in the same person or in a member of the same family. Surely most diseases and syndromes, when first discovered or investigated, equally merit the term atopic on the basis of their being 'strange'. Furthermore, when a disease is well understood and no longer strange it is no longer 'atopic'. In any case the term was pre-empted by Sir Thomas More (1551) in his invention of Utopia, no place, that is an imaginary, non-existent place.

While on the topic, TOPICAL application of a drug is well described as such because it is applied in the place where it is effective, rather than given systemically to be transported to the desired place by the blood stream. We used to say LOCAL (L. *locus* place) and we still speak of local application and local anaesthetic, but that term stuck before it was generally agreed that three syllables are better than two. And what about TOPICS for discussion, and topic teaching? Starting from the title of a work by Aristotle, *Ta Topika* (Concerning Commonplaces) considerations common to many kinds of subjects, thence by a long, tortuous and barely recognizable route to its present meaning.

• Anglo-Saxon and Graeco-Roman derived anatomical adjectives

For many of the common parts of the body there is a choice of adjectives, which may be of either Graeco-Roman or Anglo-Saxon origin. In each instance there is an important difference in meaning between the two adjectives pertaining to the relevant part, and it is the Graeco-Roman adjective that conveys the direct meaning. For example, the adjective of hand is manual, not handy; of tooth, it is dental not toothy; of throat it is pharyngeal, not throaty; and of leg crural, not leggy. It is curious that the corresponding Anglo-Saxon-derived adjective is usually pejorative and often denotes a feature of the personality of the owner. Similarly, the adjective of nose is nasal, not nosey; of heart, cardiac, not hearty; of head, cranial, not heady; of cheek, malar or buccal, not cheeky; of skin, dermal not skinny; of blood, haemic or haematological (depending on context) not bloody; and of cock, the adjective is penile, not cocky. Alas, more pitfalls for foreigners. A difficult language, English.

• Homophones and homonyms

HOMOPHONES are words which sound the same but have different meanings; some are even spelt the same (HOMONYMS). The context usually conveys the right sense and confusion rarely occurs. We know that men's underclothing is not being discussed if someone says, 'His breath came in thick short pants'.

But take, for example, psychosis and sycosis. Which one is meant? Why! A mental (p)syc(h)osis of course. Mental(chin) or mental(mind)?

Pleural and plural. Secretary: 'Did you say lobe or lobes?' Pedantic doctor: 'I said the plural'. Difficulty has also been experienced with 'inattention' and 'inner tension'; and 'pharmacology' has appeared as 'farm ecology'. (Shorthand, too, has its homographs. In the Pitman system the outlines for apices and bases are identical—opposite meanings!)

Natal may refer to birth or buttocks. Fissure and Fisher. Heel and heal. 'Is the liver friable?' means one thing in the post-mortem room and another (one hopes) in the kitchen.

Genuine confusion can occur in England but not in Scotland between the spoken oral and aural. South of the border both sound like 'awrel'; north of it they are often pronounced respectively 'orral', with a short o, and something like 'owral'. Sassenachs, please copy. Alternatively, why not adopt the Greek roots, stomal and otic?

One near-homophone is every doctor's nightmare, namely the close similarity, especially during rapid speech, between hyper- and hypo-. It is just too bad that they have opposite meanings. When dictating letters, a doctor is at the mercy of any secretary who happens to be no more than a mindless typist. Here too the situation could be avoided by adoption of the Latin roots super- and sub-. Does supertension sound strange? You would get used to it, and in three months forget you ever said hypertension. And anyway, hypertension has a hybrid derivation (as if you cared!).

• Scientists and atoms

Some commonplace words date from antiquity but their meanings have changed by drift interspersed with occasional jumps. SCIENCE is an old word (1340); it meant 'the state of knowing' (L. *scio*, *sciens*, knowing). In the Middle Ages the Seven Liberal Sciences comprised a lower division, the Trivium (grammar, logic, and rhetoric), and an upper division, the Quadrivium (arithmetic, music, geometry, and astronomy). The trivium has yielded TRIVIAL in biological nomenclature, (a) in vernacular usage for ease of communication, and (b) more strictly, the specific as opposed to the generic name; also, in a general sense, familiar and trite. I think it is well on the way to being equated with 'trifle', a matter of little value or importance, with which it has no connection. But, to return to Science. As we understand the word today it was originally known as natural philosophy, and a scientist was a natural philosopher. The term science was not adopted in place of natural philosophy until early in the nineteenth century. A person well versed in that discipline was known as a man of science; SCIENTIST was tentatively suggested in place of this somewhat cumbersome expression in 1834 at a meeting of the British Association for the Advancement of Science. Scientist has stuck, and is suitably free from identifiable gender. Not so in German, with *Naturwissenschaftler* plus *-in*, if it is a lady.

ATOM (Gk. *a-*, not; *tomos*, cut/table) is another word revived from antiquity; it means, a particle so small as to be incapable of further division. The concept that matter is made up of indivisible particles goes back to Democritus (*c*.460–357 BC) and was elaborated by Lucretius (*c*.98–55 BC). It was Dalton (1766–1844, famous for his original work on colour-blindness) who based his atomic theory on the observation that in chemical compounds the elements always united in definite proportions. It was he who revived the term atom in its modern sense.

• Primary, idiopathic, essential, and cryptogenic

The use of these terms betrays ignorance of the causes of the conditions to which they are applied. The use of the first three made sense in the days before scientific medicine. Later, they were a useful convention as a screen to hide one's uncertainty regarding aetiology. After a decline in usage in the third quarter of this century, they are now becoming fashionable again.

Primary, essential, and idiopathic, when applied to diseases, all meant originally that the disease or condition existed in its own right and was not due to any other disease or condition. Since this concept of aetiology is no longer acceptable, these terms are currently applied when the cause is unknown. When this is the case, we should say so, and CRYPTOGENIC (Gk. *kryptos*, hidden; *genesis*, origin), as in the accepted term cryptogenic fibrosing alveolitis, is a perfectly good way of doing this. So, I say: up with cryptogenic and down with primary, essential, and idiopathic!

PRIMARY was used in contradistinction to secondary. There has been no corresponding recrudescence of secondary in describing conditions labelled as primary. To take an example; in the 1930s anaemias were classified as secondary if due to bleeding or dietary lack of iron, and as primary if the diagnosis was a total mystery, as in pernicious anaemia or acholuric jaundice. In the 1940s we had primary atypical pneumonia. In both instances the term primary was dropped as soon as the causes were discovered.

ESSENTIAL, as in hypertension and thrombocytopenic purpura, does not mean one must try to get it regardless of cost. Essence is not easy to define in this context. It is that by which a thing is what it is and by which it differs from other things. A definition of essential which comes close to this usage is given in the *OED* (and marked obsolete) thus: dependent on the intrinsic character or condition of anything (—not on extraneous circumstances). It entered modern English from the German of E. Frank[1] in 1911; he described hypertension not due to renal disease as *essentielle Hypertonie*.

IDIOPATHIC was for many years limited to little beyond epilepsy, but its applications have burgeoned and I counted eleven in recent journals. The IDIO root derives from the Greek *idios*, meaning own, personal, private. IDIOSYNCRASY is a personal, individual response, not the usual or expected one, to a drug, food or other substance or influence. *Idiōtēs* was not a pejorative term; it denoted a private person in the sense of not being a public servant. The meaning shifted (as meanings do) to indicate a person with no special knowledge; thence, an ignorant, uncouth person, whence IDIOT eventually came to mean a person so deficient mentally as to be incapable of ordinary reasoning or rational conduct. In modern Greek, *idiōtikos* still means private (Fig. 6).

Fig. 6 Sign outside a private house in Cyprus: 'Idiōtikos khōros' (Private place)

[1] Frank, E. (1911) Bestehen Beziehungen zwischen chromaffinem System und der chronischen Hypertonie des Menschen. *Deutsches Archiv für Klinische Medizin* **103**, 397.

• Cell

The CELL is the fundamental unit of the body's architecture, and indeed of all plant and animal structures. Even tissues lacking a cellular structure, such as fibrous tissue, bone, and hair, have their origin in cells. The word is derived from L. *cella*, a small room, especially one of several in the same building, for stores, for a monk, hermit, slave, or prisoner. Many homes still have a cell for purposes of storage; it is the cellar. It has been suggested that the original *cella* was that of the honeycomb (L. *cera*, wax). There are countless instances of shifts of the letter *r* to *l*, and of *l* to *r*, in writing and speech within the same word, and vice versa, within and between languages, so *cera* to *cella* is a feasible one. The first appearance of 'cell' in the literature of the biological sciences was in the field of botany (1665)[1], about 55 years after the invention of the microscope. The similarity of plant cell walls seen in section to a group of small rooms seen in the plan of a building made it a good choice for this newly-discovered structure. (It is, incidentally, the material of the plant cell wall that gives us CELLULOSE, celluloid, Cellophane, and Sellotape.) From the cell wall to the inclusion of its contents was a small semantic step. In the zoological sciences the cell wall has thinned to a membrane, and even this has merely the structural flimsiness of a physicochemical interface. The cell today is its contents, membrane and all, as a unitary structure.

Brain cells are different—at least they used to be. From the fourteenth century onwards it was supposed that they were cavities or, as it were, pigeon-holes in the brain which were the seat of various mental faculties and where knowledge was stored. The *OED* quotes from Matthew Prior's poem 'Alma, or The Progress of the Mind' (1720),

> The brain contains ten thousand cells,
> In each some active fancy dwells.

The modern brain has rather more. One needs a functional reserve if it is true, as has been alleged, that one loses one hundred thousand a day during normal ageing.

AREOLAR tissue (L. *area*; *areola*, a little area) is so called because of the small spaces between the connective tissue fibres.

These spaces were likened to cells in the original sense of L. *cella*, hence cellular tissue, an older, now outmoded, term for areolar tissue. When the tissue is inflamed we still speak of CELLULITIS. There is another AREOLA (little area) whose possible function is that of the target's inner ring to guide the hungry infant's lips to the bull's-eye.

[1] Hooke, R. (1665) *Micrographia*, pp. 113, 116. J. Martyn and J. Allestry, London.

• Cataract

CATARACT (Gk. *kataraktēs*, down rushing) in its non-medical sense means a large waterfall, especially one falling over a precipice, as opposed to a cascade, which is one of a series of small waterfalls. In medical terminology a cataract is an opacity of the crystalline lens of the eye or of its capsule, or both. An obsolete usage of cataract was for portcullis (F. *porte coulisse*, sliding door) derived from its downward fall, and it is from here that the medical term was adopted. Just as a portcullis shuts off entry to the castle, so a cataract occludes entry of light to the eye. The French surgeon Ambroise Paré used both terms, *cataracte ou coulisse* (*c*.1550). Immunologists describe as a CASCADE the sequential changes occurring in the activated complement system.

• Star

Despite the fact that the stars at night present point sources of light, the word 'star' has been extended to describe a geometrical figure of radiating points, or a number of rays

Fig. 7 Detail from the title-page of *Introductio Geographica Petri Apiani in Doctissimus Verneri Annotationes* (1533) by Petrus Apianus, (showing a method of measuring longitude). The stars are depicted with radiant points.

diverging from a central area (Fig. 7). Indeed the term 'star-shaped' is used to describe such a geometrical figure (rather than the sphere we know it to be). Hence the STELLATE ganglion and the DIASTER in mitosis. Presumably this extension of meaning arose from the fact that bright stars at night are not seen as points. Small refractive errors, due to the distribution of fibres in the lens of the eye, cause them to appear with small spikey projections. Two or three such projections in the image perceived by each eye would yield up to six rays. I offer this explanation as the *OED* is silent on this derived usage.

Patrick Trevor-Roper, writing in *World Medicine* about cataract (14 July 1979, p. 20) refers to starry-eyed oculists. As

Star is German for cataract, I wonder whether this was a chance juxtaposition or a deliberate EEC pun?

• The chancellor at the gate

CANCER is Latin for crab, and came to be applied, as CANKER and CHANCRE, to chronic ulcerating lesions generally. In the past one or two hundred years it has come to mean more specifically a malignant tumour. This may also have been the meaning in ancient times because, according to Galen, it was a tumour and called cancer because the surrounding veins resembled a crab's legs. The Greek for crab is *karkinos*, whence CARCINOMA.

The word cancer had also an unrelated meaning, namely, grating, and was commonly used in its diminutive plural, *cancelli* (Fig. 8), meaning cross-bars, grating, and lattice (*Cancelli* is modern Italian for a barred gate; see Fig. 9). It has yielded some English words which we hear from time to time in clinical and academic practice. A patient may CANCEL an appointment. Indeed, I depend on cancellations to finish the morning clinic in time for lunch. To cancel, originally, was

Fig. 8 *Cancelli* (Latin): Roman lattice-work. (From *Orbis Pictus Latinus* by H. Koller, 1983. Courtesy of Artemis Verlag, Zurich.)

Fig. 9 *Cancelli* (Italian): a barred gate.

to obliterate writing by drawing lines across in a lattice. A single line or series of parallel lines drawn across would be only half the job, etymologically at least.

CANCELLOUS BONE is constructed of thin partititions enclosing cavities containing red marrow. Where the forces acting on the bone are predominantly unidirectional, the partititons are aligned in the direction of the applied force. The cut surface of cancellous bone then shows the partitions arranged in two series of parallel lines crossing each other as in a lattice (Fig. 10).

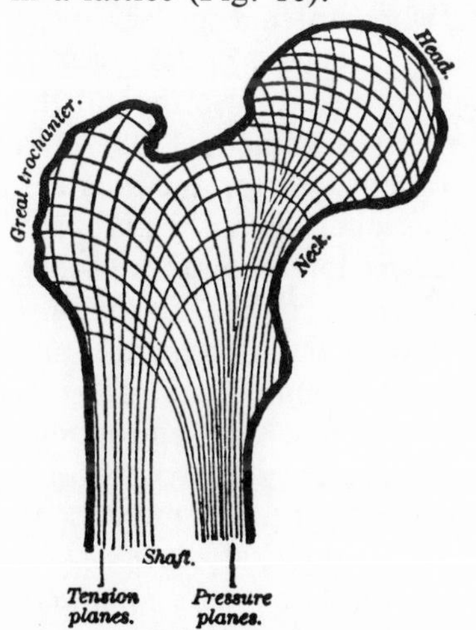

Fig. 10 Diagram showing the arrangement of cancelli in the cancellous bone in the neck of the femur. (From *Gray's Anatomy*, 15th edn, 1901.)

A chancel was originally a lattice-work screen in a basilica separating judges and council from the audience. In a church it is no longer the screen but the eastern end, reserved for clergy and choir, beyond the screen. A CHANCELLOR was originally a court usher whose station was at the grating (chancel) which separated judge from public. His promotion over the centuries has been meteoric. He left the grating and became an official secretary, then the secretary of a nobleman. Eventually enobled himself, as Lord Chancellor he has become the highest officer of the Crown, and has some quite important duties, such as Keeper of HM Conscience. By extension therefrom, the titular head of a university is its chancellor. The vice-chancellor does most of the work and, when times are difficult, the grating occurs with his teeth.

• Chiropodist

Originally a chiropodist was one who treated disorders of the hands and feet (Gk. *cheir*, hand; *pous*, gen. *podos*, foot; a hand-foot-ist), but he was soon confined to disorders of the feet alone. Apparently the term was coined (1785) in a pretentious attempt to bestow exalted status on the practice, an attempt which drew the following comment in *The European Magazine* of the same year; 'Classic lore has now reached Davies Street, the residence of our Lucretian Chiropodist (anglice corn-cutter). But whilst we point out the absurdity and needless affectation of learning, the coining of new-fangled derivatives on every occasion, . . . ', (*OED*). Well, chiropodist has come to stay. The Americans prefer 'podiatrist' (foot-healer), which avoids the irrelevance of the first moiety. As to the pronunciation of chiropodist, several versions are current. The *OED* favours kye-ropodist; *Chamber's Twentieth*

Century Dictionary, ki-ropodist, with shi-ropodist as second choice. Even tshi-ropodist is heard from some who seek relief from painful feet.

• Coughing, huffing, and ahem

Everyone knows what a cough is. The huff is perhaps less well known. It is a forced expiration without prior glottal closure. Huffing is advised in the Brompton Hospital booklet on physiotherapy for the chest as an aid to expectoration. As one with a special interest in chest diseases, I can testify to the frequent efficacy of huffing when coughing has failed to shift a sticky pellet of mucus.

The word COUGH is clearly echoic. The c mimics the opening of the glottis with sudden release of pressure of air from within the chest; the f (phonetic equivalent of gh) mimics the sound of air travelling at speed in the trachea. Terminologically, a huff can be regarded as a cough which lacks the glottal c.

Let us see whether our fellow Europeans are coughers, linguistically speaking, or huffers, and let us locate the glottal component as represented in speech. Beginning anteriorly with the ancient Greek bilabial, we have *bēx*. Modern Greek places the glottal plosive equivalent a few millimetres further back with a labio-dental 'veekhuss'. The Romans were just behind with *tuss/is*; (you can almost hear a faint wheeze in the double s). The consequential Romance languages also mimicked the glottal plosive with t – It. *tosse*, Sp. *tos*, Port. *tosse*, F. *toux*. Further back still with k are the Slavonic languages, e.g. Czech, *kašel*, (kushell), and Dutch *kuch*, and Hungarian *köhög*. The Teutons seem to be huffers; thus, Ger. *Hust/en*, Dutch *hoest* (hoost), Dan. *hoste*, Sw. *hosta*. But these last mean cough, not huff. The Germans, like the Dutch, also have a cough-type word in *keuchen*, to pant, gasp, wheeze, and whooping cough is *Keuchhusten*. The German *Hauch*, breath (in the sense of exhalation) comes close to huff, but lacks the expulsive force of the English word. There seem to be no words in other languages corresponding to huff in a clinical context. I am told that the Eskimos have twenty words for snow. Perhaps, as a nation of expectorators (before the Clean Air Act) it is appropriate that we British should have a verbal repertoire in that area of activity.

Finally, there is AHEM, thus spelt to indicate a 'sub-cough', as it were; a throat-clearing action. Here we have retreated anatomically to the glottis.

• Infarct

What is an INFARCT? The term denotes ischaemic necrosis. There is an implication that only part of the involved organ is necrotic, for when the whole organ or limb has undergone ischaemic necrosis it is customary to apply the term gangrene. By derivation, infarct means stuffed (L. *infarcire*, to stuff; p.p. *infarctus*). This is true anatomically only for lung infarcts where, because of the double blood supply, the alveoli are full of blood and the lung parenchyma is consolidated. Actually, the past participle of L. *farcire* is *fartus*. I imagine that this was too near the wind for its adoption by even the most morbid of anatomists. It seems the Romans had difficulty in pronouncing -ct- and dropped the c as their Italian descendants have done; for example, with arctic, in Italian *artico*.

In a myocardial infarct the myocardium is, on the contrary, ischaemic. In culinary French *farci* means stuffed, whence the English force, as in forcemeat, which is stuffing made from chopped seasoned meat. Etymologically, the only true cardiac infarcts are stuffed hearts for eating, the *coeurs farcis* of the chef, not the *infarctus du myocarde* of the French physician.

FARCY or glanders is an infectious disease primarily of horses, occasionally communicated to man. In this condition subcutaneous nodules appear along the lymphatics, which, because of their stuffed appearance, are known to the laity as farcy buds. And what is a FARCE but a play, originally short, whose object is to excite laughter? This derives from the interludes of impromptu buffoonery which actors in religious dramas were accustomed to interpolate or, as it were, stuff between the acts.

• Dollars and lung cancer

The Erzgebirge (literally Ore-mountains) are a 90-mile-long range stretching along the north-western frontier of Bohemia in Czechoslovakia, where it adjoins Saxony in East Germany. It is an area famous for its health resorts and for the mining of various metallic ores. On the Bohemian side lies the Joachimsthal (Joachim's Dale). I suppose most doctors know of the high mortality from bronchial carcinoma affecting miners in the Joachimsthal, which is attributed to the inhalation of radioactive particles, possibly aggravated by traces of arsenic. But did you know that the American dollar was named after the joachimsthaler? This was the standard German monetary unit for 350 years until superseded by the mark in 1873; it was so called because it was minted from silver mined in the Joachimsthal. Joachimsthaler was too long a word, even for Germans; it was shortened to 'thaler' (pronounced 'tahler') whence the word dollar (pronounced 'dahler') derives.

Calligraphy and cacography

CALLIGRAPHY (Gk. *kallos*, beautiful; *graphos*, writing) is beautiful writing, elegant penmanship; it is a well-known word even to doctors. CACOGRAPHY (Gk. *kakkos*, bad; *graphos*, writing), is a less well-known, indeed almost unknown, word and one that deserves to be better known. The reputation of the medical profession for cacography is doubtless well merited. In my experience the perpetrators are predominantly clinicians, that is, those who write at least part of the time at the bedside. Why do doctors in general write so badly in shaping letters and words? I suggest it may be acquired as a habit when working under pressure. Yet, unless one is writing a memo for one's own reading, cacography is an act of thoughtless discourtesy towards the reader. (Just another stone from the glasshouse!)

There are many stories about doctors' handwriting. One of the best tells of a young lad in need of a reference for a job, the referee to be a person of responsibility and repute in the locality. 'Well', said his dad, 'go round to Dr Brown. He's known you all your life.' The boy returns from Dr Brown with a handwritten testimonial. 'Let's see what he's written,' says Dad. The envelope is steamed open; the note is perused, but the message is illegible. 'Go round to Jones the chemist; he's used to doctors' handwriting,' says Dad. The boy does as he is bid and hands the note to the chemist who, after a preliminary glance, retires to a back-room to submit it to detailed scrutiny. The chemist returns five minutes later, and places a bottle of tablets on the counter with the words, 'That will be 70p!'

Capillary

Originally an adjective meaning to have a very small, hair-like internal diameter, thus describing a tube or one of the very slender blood-vessels connecting arterioles and venules (for instance, capillary tube, vessel) CAPILLARY is now commonly used as a noun. Capillary, hair-like, is from L. *capillus*, a hair; not any hair, but a hair of the head (L. *caput*, head; *pilus*, hair). Robert Hooper's *Lexicon Medicum* (1839) describes ten different names for hair according to location. It is not clear why scalp hairs should have been the exemplar for capillary blood vessels or tubes. I suggest that the choice depended on their greater length, (see *Hair*).

• Carbuncle, anthrax, boils, and bullae

The cardinal signs of inflammation, according to Celsus (AD 30), are L. *calor*, *rubor*, *tumor*, and *dolor*—heat, redness, swelling, and pain. CARBUNCLE (L. *carbunculus*, a small glowing coal) manifests the heat and redness. *Carbunculus* is a diminutive of *carbo*, coal, burning or burnt wood, whence also the element CARBON. ANTHRAX, a severe infection, usually of the skin in humans, similarly is Greek for coal; that is, taken as glowing in this clinical context. An accumulation of coal-dust in the lungs, seen in miners, is ANTHRACOSIS. Here come more pitfals for foreigners! In French *le charbon* means anthrax (or coal), while *l'anthrax* means boils. BOIL, on the other hand, does not denote the *rubor* of inflammation, but the *calor* and *tumor* in the sense of ebullition, that is, bubbling from the application of heat. The Latin for bubble is *bulla*, and BULLA in morbid anatomy well describes an air-containing vesicle in the lung, often seen in emphysema, and a large fluid-containing blister in the skin. FURUNCLE, a boil, and FURUNCULOSIS, the state of multiple boils, owe their etymology to an alleged similarity between the swellings in the skin and a disease of vines, in which there are knobs on the plant which 'steal the sap'. L. *furunculus* is a petty thief, diminutive of *fur*, a thief, whence we have furtive, in the manner of a thief.

• Candidate

In the professions a candidate is one who has entered for an examination, or who has applied for a job in the face of competition. L. *candidus*, white, refers not to the colour of the applicant's face at viva voce examination or at interview, but to the white togas worn in ancient Rome by applicants for office. Today it still helps to be appropriately dressed when attending for interview, and to exhibit a CANDID (frank and open) demeanour.

• Cocks and taps

A cock is a male domestic fowl and, by extension, the male of other bird species. The aggressively self-confident attitude of a fighting cock gives us 'cocky'. Over-confident is 'cock-sure'. 'Cock of the walk' is one who will not tolerate a challenge to his supremacy; hence, 'Cock!' is a familiar greeting between men implying recognition of *machismo*.

A cock is also a short spout or pipe serving as a channel for passing liquids and having a device for regulating or stopping the flow. The derivation of this sense is obscure. The *OED* says 'the resemblance of some stop-cocks to a cock's head with its comb readily suggests itself' (Fig. 11). A coxcomb-like

Fig. 11 Ancient zoomorphic stop-cock with stylized valve-handle in the form of a male domestic fowl. (From *Orbis Pictus Latinus* by H. Koller, 1983. Courtesy of Artemis Verlag, Zurich.)

structure is sometimes found on spouts to hold the handle of a pail.

A cock in firearms is a spring-loaded lever holding a match, flint or hammer, and capable of being raised and then brought down by the trigger, thus igniting the powder; it is so called because it was appropriately zoomorphic in its original form.

A cock, in vulgar usage, is a penis. The term in this sense derives from the pipe, as described above. Although the *OED* gives 1730 as its first recorded usage, oblique earlier reference to the ithyphallic state suggests an earlier date, possibly in unrecorded speech. Shakespeare surely intended a *double entendre* (as in the later tradition of Restoration plays and music-hall) when he wrote (1598) '. . . and Pistol's cock is up and flashing fire will follow' (*Henry V*, V, ii, 54). Furthermore, the ithyphallic principle seems to underlie the verb 'to cock'. Although the *OED*, in defining it 'to stick stiffly up or out', likens this posture to a cock's neck in crowing, this suggestion would appear to come from a lexicographer with a Freudian scotoma. 'Cock up' is applied to appropriately-shaped articles, as, for example, the plaster of Paris cock-up splint.

The use of 'dicky' for penis in children's language derives from rhyming slang for cock, and is the first moiety of Dickory Dock.[1] Genital terms are vulgarly used as vehicles for insult. 'Cock' is thus used to describe something nonsensical, though it is probably of other provenance, (? poppycock, Du. *pappekak*, porridgy shit; cock-and-bull story). In slang, a cock-up is an action that has ended disastrously.

It is surprising that in German the word *Hahn* has all the same meanings as in English: male bird, tap, firearm component and penis (also the diminutive *Hähnchen*). This parallelism of meanings is quite remarkable and would seem to be beyond mere coincidence. Albrecht Dürer's pictorial pun in the woodcut of 1498, *Männerbad* (The Men's Bath) contrives to show three of these meanings in one small area (Fig. 12); the male bird is represented on the tap-handle. The spout of the stop-cock is so placed as to hide discreetly the man's genital area, though in position, size and shape it well-represents the penis. As Dürer is reported to have said, 'Det voss my liddel dschoke'[2].

A TAP, on the other hand, was originally a peg, plug, or bung for opening and closing a hole in a cask or similar vessel;

Fig. 12 Männerbad (men's bath)—detail. (Albrecht Dürer, 1498.)

later it became a hollow plug in a containing pipe for regulating or stopping the flow. To tap is to pierce the wall of a vessel to draw off the liquid contents, as tapping an ascites or pleural effusion. The corresponding German word for bung is *Zapf* (a German z often became English t on crossing the North Sea). The original meaning of *Zapf* as peg-shaped has been retained in German; its diminutive *zäpfchen* means both suppository and uvula. In English, we liken the latter to a small grape (L. *uva*, *uvula*).

TAMPON (Old F. *tapon*, from Germanic) is a plug to stop bleeding or absorb secretions. The military equivalent of the publican's call to his customers announcing 'Closing time!' is a drumbeat or bugle call, the TATTOO, from seventeenth-century 'tap-too', meaning 'tap shut', adopted from the Dutch *taptoe*; Ger. *Zapfenstreich*.

> Thomas the Earl of Surrey, and himself,
> Much about cock-shut time, from troop to troop
> Went through the army, cheering up the soldiers.
>
> (*Richard III*, V. iii. 70)

(The other tattoo, a mark or design on the skin made by insertion of pigment, is a Polynesian word.) Strictly speaking, then, a cock was a pipe and a tap was a bung.

The cock's crow has traditionally been a wakener for early risers. As the Hungarian diplomat is reported to have said at an embassy dinner, 'My speech will not be long. As you English say: Early to bed and up with the cock!'[3]

[1] Franklyn, L. (1961). *A dictionary of rhyming slang* (2nd edn). Routledge & Kegan Paul, London.

[2] Dürer, A. (1498). *Zeitschr. f. Kunst u. Wissenschaftl. Unsinn.*

[3] Blake, G. F. (1978). *Book of bricks* (ed. R. Morley) p. 123. Weidenfeld & Nicholson, London.

• Tribadism

Anyone not knowing better might be pardoned for thinking that tri-bad-ism means 'three bad girls', whereas in fact it means 'two bad girls'. The trouble is Gk. *tribos*, rubbing, and tribadism means (as if you didn't know) the lesbian practice of rubbing the external genitals against those of the partner. By the same token TRIBOLOGY is not the study of tribes; it is the study of friction in relation to biological surfaces, notably those of joints.

• Limb, limbus, limbo

There are two sorts of limb: (a) a leg, arm, or wing (of obscure etymology), and (b) (L. *limbus*) an edge or border in a heavenly body (astronomically speaking) and in anatomy as, for example, the margin of the cornea and of the fossa ovalis in the right atrium of the heart. LIMBO, ablative of *limbus*, from 'in limbo', is, in the theological topography of Catholic belief, a region on the border of Hell, and the abode of unbaptized infants who remain suspended indefinitely neither in the One Place nor in the Other; by metaphorical extension, it refers to analogous situations in this world. Hence, to be 'out on a limb' does not mean 'held at arm's length'; it means to be on the periphery, or outside the desired area of personal involvement.

• Deflatulate

DEFLATULATE is not in any dictionary—yet. As is the case with several bodily functions, the descriptive term of Anglo-Saxon origin is monosyllabic, four-lettered, and taboo in what may be linguistically called 'polite society', yet freely used and perhaps freely performed within the family circle and the company of intimate friends and comrades; for example, FART (O. E. *feortan*; Ger. verb *furzen*, noun *Furz*; to pass intestinal gas through the anus). Most of the bodily functions can be described by single words suited to polite society or physiological terminology—for example, eructation, mastication, sternutation, micturition (or preferably miction), defaecation, copulation. But there is no single corresponding word for 'fart'. Doctors often experience difficulty in making their meaning clear to patients when enquiring about this aspect of bowel function, and both parties tend to use circumlocutory phrases which are capable of being misunderstood.

WIND is dangerously ambiguous; many patients use this word when they mean abdominal pain. 'The passage of wind' evokes the time-wasting question, 'Up or down?' 'The passage of wind from the back passage' is understood by any patient but is too cumbersome. 'Crepitation' was used in the nineteenth century and referred to the acoustic properties of the act. This word is now used for sounds arising in the lungs, bones, joints, or subcutaneous tissues. In my lecture to the Listerian Society at King's College Hospital in 1974, I proposed that the act of passing rectal flatus be termed DEFLATULATION in circumstances unsuited to the monosyllabic alternative. This single word is self-explanatory, and I suggest it be adopted for medical, and perhaps general, usage. Flatulate has been used by the singer-guitarist Jake Thackray;[1] this has the merit of simplicity and of beginning with the familiar f. Pedition (L. *pēdo*, p.p. *pēditum*), and its notional verb pedite, have been suggested, but could be confused phonetically with petition, expedite, and expedition. The study of intestinal gas formation has been termed flatology,[2] but we are concerned with a suitable word for its expulsion *per anum*. Downwind[3] says what it means and, coined from Anglo-Saxon roots, it may be suitable for the generality of patients. Incidentally, raspberry is the first moiety of raspberry tart in rhyming slang for fart,[4] but it refers to the mimetic effect created by blowing between constricted lips.

Chaucer's use of fart suggests that in his day the word could be used without embarrassment.

> 'This Nicholas anon let flee a fart
> As greet as it had been a thonder-dent (-clap)'

and again, writing about the elegant parish clerk, Absolon,

> 'He was somdel squaymous (somewhat squeamish) of farting'.[5]

The Elizabethans used 'ventosyte' (ventosity) to describe either a state of gaseous distension, or the gas-forming potential of certain foods.[6] In the latter sense, VENTOSITY is a word that merits revival. By what seems an unfortunate choice of word, the slit in the lower edge of a jacket at the rear is known in tailoring as the vent.

Many people believe that retention of rectal flatus is harmful. An epitaph in Leyland churchyard, Lancashire, says so in verse.[7]

> Let the wind go free
> Where'er thou be
> For 'twas the wind
> That killéd me.

A clear case of ileus.

[1] Thackray, J. (1977). *Jake's progress*, p. 44. W. H. Allen, London.
[2] Levitt, M. D. (1980) Flatology. *New Engl. J. Med* **302**, 1474–5.
[3] Anon. (1983). Downwind with sweeteners. *Drugs & Therap. Bull* **21**, 104.
[4] Franklyn, J. (1961) *A dictionary of rhyming slang* (2nd edn). Routledge & Kegan Paul, London.
[5] Chaucer, G. (*c*.1387) *The Canterbury Tales* (The Miller's Tale).
[6] O'Hara-May, J. (1977) *Elizabethan Dyetary of Health*. Coronado Press, Lawrence, Kansas.
[7] Silcock, A. (1958, 1977). *Verse and worse, p.125. Faber & Faber, London.*

• Faex

On a previous page I refer to the lack of single words that may be used in polite society, or in a medical or physiological context, to describe certain bodily functions (see *Deflatulate*). All are agreed that an accurate description of faeces may be of clinical importance. While there is no difficulty in describing consistency, colour, or abnormal constituents, when it comes to mentioning a single more or less cylindrical portion, we are stuck for a word. We are apt to flounder with vague or circumlocutory expressions that are capable of being misunderstood by patients. 'A formed stool' refers to an entire evacuation, not necessarily to a portion thereof. In *The Canterbury Tales*, Chaucer's own contribution (Sir Topaz) is interrupted in mid-sentence by the Host because he is bored by all that rhymed verse. To quote Nevill Coghill's modern version,[1]

> 'My ears are aching from your frowsty story!
> The devil take such rhymes! They're purgatory!
> That must be what's called doggerel-rhyme,' said he.
> 'Why so?', said I. 'Why should you hinder me?
> In telling my tale more than any other man,
> Since I am giving you the best I can?'
> 'By God,' said he, 'put plainly in a word,
> Your dreary rhyming isn't worth a turd!'

No possibility of misunderstanding here. Well, to put it bluntly and stop beating about the bed-pan, we need an acceptable word for turd.

Vincent J. Derbes,[2] writing in the *Lancet* (1957, **1**, 994), drew attention to this need and suggested a remedy. As he put it clearly and elegantly, 'There are, in the vernacular, two words which approximate in meaning to the commonly used faeces. One is used exclusively as a substantive to denote a discrete mass of faecal matter, the other as a verb or noun. Their persistence through the centuries attests as well to their utilitarian as to their scatological value. Physicians need an acceptable word which will permit the greater flexibility inherent in our Anglo-Saxon heritage. Faeces, a collective noun in plural form, lacks precise meaning. It is the singular which we require; this will permit us to say, in one word, that unit of a formed stool delimited by two successive anal sphincteric contractions. Harper's *Latin Dictionary* (1907, p. 720) gives *faex*, *faecis* as this singular. Some clumsy circumlocutions might be avoided by incorporating faex into our medical vocabulary.'

Words meaning single portions of excrement exist in the biological sciences for the product of small mammals, namely, pellets (which implies a more or less spherical shape) and droppings, usually in the plural because they are usually multiple. Pellet and dropping are words that have other meanings in differing contexts, but confusion is not likely to occur. Even so, it is preferable to have a term that can mean one thing only, and specific for humans, as, for example, fumets from deer. R. A. Davis, writing from the Ministry of Agriculture, Fisheries and Food (1979), informs me that scat (Gk. *skōr*, gen. *skatos*, dung) is used in American literature. He suggests, as an alternative, sterk (L. *stercus*, dung). Before making a choice, perhaps we should wait a while. The 'permissive age' in behaviour of the 1960s was followed by the permissive age in speech, and words that were previously unprintable are now bandied about on television and seen in the newspapers.

[1] Chaucer, G. *The Canterbury Tales*. Translated by Nevill Coghill, 1951, 1958, 1960. Penguin Classics, London. Reprinted by permission of Penguin Books Ltd.

[2] Derbes, V. J. (1957). *Lancet*; **1**, 994.

• Mnemonics for amnesia

AMNESIA (Gk. *amnēsia*; *a-*, privative; *mnēmōn*, mindful; *mnasthai*, to remember) is a failure to remember, or loss of memory, especially in a medical context if it occurs to an abnormal degree. ANAMNESIS is the word used in many European languages, also rarely in English, for a medical history Gk. *anamnēsis*, remembrance; *ana*, back; with *mnēmōn*). An ANAMNESTIC REACTION is one in which there is augmented production of antibody after a previous response to the same antigen, the immune system remembering, as it were, the previous episode. AMNESTY is a ruling authority's intentional forgetting, or overlooking, or pardon, for past offences; it is sometimes called an Act of Oblivion. Most of us have been helped during academic studies to memorize a seemingly arbitrary list of facts by the device known as a MNEMONIC. I wonder if Timothy still Doth Vex All Very Nervous House-maids (the mnemonic I used for the order of tendons, vessels, and nerve lying beneath the flexor retinaculum at the ankle). If so, he must be having some difficulty in locating someone in this probably obsolete occupation. In the days when this mnemonic was coined, every medical student's parental household must surely have had at least one very nervous housemaid.

• Rheum and rheum

These two words share the same spelling but differ in pronunciation, meaning, and origin.

RHEUM (1) pronounced room, from Gk. *rheuma*, flow, stream, is a watery or mucous secretion or exudate, especially one from the eyes or nose.

RHEUM (2) pronounced ree-um, is modern Latin for rhubarb.

Rheum (1) is obsolete. It persists vestigially in French as *rhume de cerveau*, cold in the head (see *Phlegm*) and as *s'enrhumer*, to catch a cold. In English, 'rheum' took a curious turn in the late seventeenth century, in its adjectival form, to RHEUMATIC, apparently in the belief that the painful swelling was due to a downflow ('defluxion') of humour to the affected part. A further shift brought 'rheumatic' to describe any painful condition of the locomotor system, especially if aggravated by movement.

RHUBARB is a plant whose rhizome contains an anthraquinone purgative (*Rheum palmatum*), whose leafstalks are eaten as a dessert (*Rheum rhaponticum*), and whose leaves are poisonous. Medicinal rhubarb was formerly imported into Europe from China via Russia, which explains its etymology. Rhubarb derives from medieval Latin *rhabarbarum*; *Rha* (from the Greek) was the ancient name for the Volga, and *barbarum* meant foreign.

It is one thing to be able to understand the written word of a foreign language; it is another to be able to speak it. Still more difficult is comprehension when being addressed by a foreigner in his native tongue. Most difficult of all is following a lively conversation between natives. Unless one has learnt the vernacular in the country of its origin, one will be lucky to snatch more than the odd word or phrase from the seething gabble. We call this unintelligible gabble 'Double Dutch' for self-evident reasons. Germans, when similarly confused, call it *Kauderwelsch*, from *kaudern*, to gobble like a turkey cock, and *Welsch*, a foreign language especially Romance (not, as you might think, Welsh). The French call it *baragouin*, from their inability to understand Breton, the Celtic language of Brittany (*bara*, bread; *gwin*, wine). The Dutch (I refer to single Dutch) have borrowed both these terms, and on arrival at Harwich are puzzled by what they call *koeterwaals* or *bargoens*. Doubtless similar terms exist in most languages as the situation is as old as Babel. Of course, the ancient Greeks had a word for it—*barbaros*, the bar-bar sound probably adopted in imitation of their impression of foreign speech. By extension, *barbaros* was applied by Greeks to non-Greeks and later by Romans to anyone non-Roman-non-Greek. Thus, what we call in English barbarous, was applied to persons having uncivilized manners or brutal behaviour, and of speech, being harsh or noisy. The Latin *balbus*, stammering, is possibly related. I have elsewhere referred to the interchangeability of *l* and *r* sounds. Barbary, in the sense of countries bordering the north coast of Africa, was from ancient times called Berber or Barbar by Arab geographers, when they referred to the lands west and south of Egypt, from the Arabic *barbara*, to talk noisily and confusedly. That the Greeks and Arabs should have independently coined the same word for the same thing seems too much of a coincidence, and in view of the historical connections between these nations, some mutual influence seems probable.

It is widely believed that actors in a crowd mutter 'Rhubarb' repeatedly when they are required to give an impression of conversation as a background to the main dialogue. Be that as it may, rhubarb is certainly the term used when referring to that activity. In Germany, *Rhabarber* (rhubarb) is, so I am told, actually muttered on stage to convey background conversation. Here again we have the bar-bar sound.

Various other echoic words are used, most recently 'blah, blah' and 'yak, yak' to describe prolonged chatter. *Pace* Eric Partridge, who suggests other origins,[1,2] it is tempting to suppose these are a corruption of Sp. *hablar-hablar*, to speak, and of It. *chiacchiera*, chatter. Older echoic words in this category are jabber, chatter, gabble and gibber, hence gibberish (in which branch of linguistics I hope to be offered a personal chair).

Incidentally, the feminine name Barbara, popular after the virgin saint of that name, means 'foreign', doubtless in the nicest sense of the term, exotic rather than barbarous. Adolf von Baeyer celebrated his discovery of malonylurea in 1864 by visiting a tavern frequented by artillery officers on the Day of St Barbara, who was their patron saint. Thus inspired, he performed the verbal synthesis of joining Barbara and urea to give barbituric acid.[3] At least, that is one of several stories regarding the naming of this substance and the one I find most appealing. It was not until 1903, however, that a BARBITURATE with hypnotic properties was synthesized.

[1] Partridge, E. (1969). *A dictionary of slang and unconventional English*, Vol. 1, 7th edn, p. 61. Routledge & Kegan Paul, London.
[2] Partridge, E. (1969) *A dictionary of slang and unconventional English*. Vol. 2, 7th edn, p. 1521. Routledge & Kegan Paul, London.
[3] Goodman, L. S. and Gilman, A. (1965). *The pharmacological basis of therapeutics*, 3rd edn, p. 105. Collier-Macmillan, London.

• To kill euphemistically

When death has occurred in circumstances that a coroner would describe as unnatural, one could put it bluntly and call it a killing. No blame attaches to the act of killing if it results from a cause entirely outside human control as, for example, the fall of a roadside tree which, uprooted in a storm, falls and crushes a passing motorist. The word 'kill' is acceptable in all its monosyllabic bluntness when describing such an accidental death, or when speaking of enemies, be they human, vermin, or vectors of disease. 'Kill' strikes a chill note, and euphemisms abound in situations where the killer has no aggressive sentiments and the act is performed outside the context of war, or in defence of one's self or property.

Thus a person is 'executed' if he is put to death in pursuance of a judicial sentence. (Strictly speaking, it is the sentence of death that is executed.) In certain countries the authorities 'liquidate' their political opponents. This term always calls to my mind the murderer Haig, who liquefied his victims' bodies in sulphuric acid and then poured them down a drain. Unfortunately for Haig, the gallstones of one of his victims failed to liquefy, so he was caught and himself duly liquidated. Laboratory animals are 'sacrificed'. This accords with the *OED* definition, 'the destruction or surrender of something valued . . . for the sake of something having . . . a higher or more pressing claim'. Those who believe that the deliberate ending of human life for the relief of incurable suffering is an act of kindness, call it 'euthanasia'. Animals, whose flesh is to be eaten, are 'slaughtered'. By contrast, animals who are the 'friends' or servants of man—that is, domestic pets or beasts of burden or transport are 'put down', except for Poor Pussie who is 'put to sleep'. (This is not to be confused with the phrase adopted when driving a friend to his destination, 'Where can I put you down?'.)

• Stresses and strains

STRESSES and STRAINS are terms commonly used with no clear understanding of their respective meanings. Indeed, they are often used interchangeably, or in common parlance lumped together as in the phrase 'stresses and strains', and virtually given the same meaning.

The basic meanings of stress and strain are most clearly defined in terms of mechanics, where stress is the force applied, and strain is the resulting change, whence by analogy these meanings may be extended to physiology, psychology, and clinical medicine. Thus disagreeable events or conditions in a man's occupational or domestic life are stresses, and the resulting palpitations, hypertension, anorexia, or duodenal ulcer are strains. The fact that in general the laity, apart from engineers, may fail to distinguish these terms is no ground for the medical profession to use a similarly confused terminology. That there is a realization of the present confused state is evidenced by a discussion that took place at the CIBA Symposium on The Cardiovascular, Metabolic, and Psychological Interface held at Stratford-upon-Avon in December 1978.[1] I quote;

> V. Hrubes: In the past there has been much misunderstanding of the word "stress", which is commonly used to describe both the cause and the syndrome of reactions, i.e. the response. Lately the term "stressor" has been employed for the cause.
>
> M. Lader: We have to accept that "stress" is now used mainly to describe the response. We should avoid using it for the external stimulus as well—the "stressor". However, we assume that any powerful stimulation must be a "stress", and that any "stress" response must be the result of such stimulation. We could with advantage avoid the term entirely and speak of stimulus and response.

I regard as unfortunate any attempt to introduce a new word (stressor) and misuse an old one (stress) when the entirely satisfactory words, stress and strain, already exist.

[1] Elsdon-Dew, R. W., Wink, C. A. S. and Birdwood, G. F. B. (Eds.) (1979). *The cardiovascular, metabolic and psychological interface.* International Congress and Symposium Series, No. 14, section 1A, p. 3. Royal Society of Medicine, London.

• Jog-trot

Many people go jogging these days. It is not so much a question of where or when they are jogging as what. Their livers and spleens? To jog is to shake, shake up, or give a slight push to something, as anyone knows who has retreated backwards from a bar carrying five drinks. When this now fashionable and allegedly health-giving form of locomotion was imported from America, it brought with it the American designation 'jogging' (as an intransitive verb). Before that, the few dedicated and regular practitioners of this form of exercise called it trotting, a word which truly denotes locomotion, and which derives from the same Germanic root as 'tread'.

The *OED* tells us that a trot is the gait of a quadruped, originally a horse, in which the legs move in diagonal pairs almost, that is, front near-side with rear off-side, and the word is applied to a similar gait of man between a walk and a run. I must take issue with the *OED* here. In walking, the number of feet touching the ground alternates between one and two; in running, between zero and one. A person either walks or runs; there is nothing in between. All walkers in competitive athletics know this only too well, as becoming even momentarily airborne leads to disqualification. Trotting (or jogging) is therefore a form of running.

I have nothing against jogging for those who derive pleasure from this activity or who hope to live *bis hundertzwanzig*. After all, *de gustibus non est disputandum*. I just look askance at the introduction of new uses for old words when there are entirely satisfactory ones already in existence.

• Recumbent

Every doctor knows that, when referring to the horizontal position of the body, PRONE means lying face down, and SUPINE, face up. As students, we learnt this from the analogous palm-down and palm-up positions of the hand. The laity are often ignorant of this distinction—for example, 'he lay prone on his back'. RECUMBENT is appropriate for a body lying on a surface without reference to the direction in which the front of the body is facing. A diver, jumping from a board and while in mid-air, could be prone or supine but never recumbent. A RECLINING person is horizontal or in an inclined, sloping position head-end upwards. Figurative meanings, not surprisingly, have somewhat pejorative implications; for example, prone denotes a natural inclination or tendency, as in accident prone, while supine means morally or mentally inactive when circumstances demand action. PROSTRATE means prone in token of submission, or through exhaustion, weakness or grief. Many of my male patients tell me they have succeeded in permanently avoiding this state by having their 'prostrate' removed.

• Foreign names

How should foreign names be pronounced? It is one thing if a person emigrates. He should be allowed to alter, if necessary, what would otherwise become a foreign name to something readable and pronounceable in the country of adoption. Thus the eighteenth-century German composer Händel dropped the umlaut after he came to England and became Handel. People from all countries in Europe who flocked to America altered the spelling and pronunciation of their names, or had them altered by immigration officials, to achieve compatibility with English phonetics and orthography. It is, however, entirely another matter when we pronounce or attempt to pronounce the names of foreign colleagues. Medicine is international and foreign names abound in medical publications. Should we pronounce these names as the owners of the names do, (or did)? Or should they be Anglicized? Or something in between? You may ask, 'Does it matter so long as it is understood?' I think it does. Nowadays we meet our foreign colleagues both at home and abroad with increasing frequency. In the interests of intelligibility and courtesy, not to mention avoidance of sounding ridiculous, we should certainly attempt to adopt the correct pronunciation. That this is so becomes manifest if the boot is on the other foot. Then, according to the nationality of the speaker, Smith may become Smeece; Wright becomes Vrigt; Bates, Bah-tess; Vaughan, Fowg-hun (to a German not knowing better), while foreigners conscientiously trying to say 'Hughes' run the risk of pharyngeal spasm and trismus. That danger may be mitigated by following the continental rules of pronunciation and saying Hug-hess or achieving an approximation with Yooze.

Replacing the boot on the original foot, let us take some French surnames well known in medical history. The first column below shows the name; the second, how it is usually pronounced in England; the third gives an approximately correct native pronunciation, and the fourth uses the international phonetic alphabet in the interests of accuracy. The stressed syllable is preceded by a prime dash.

Arthus	'Ahthus	Ar'tewce	ɑʀ'tys
Braille	'Brayl	'Brye (like why)	'bʀɑ:y
Marey	'Mary	Mah'ray	mɑ'ʀɛ:y
Marfan	'Mahfan	Mar'fong	mɑʀ'fɑ̃
Danlos	'Danloss	Dong'low	dɑ̃'lo:

It must be out of respect for the great man that we say Sharko and not Tsharkott.* The French do not play quite fair with Arthus, where the terminal s is sounded; yet for Camus (the 1957 Nobel prize-winner for literature) the s is silent. But, great heavens, who are we to complain?

Pardonable difficulties arise with Lutembacher, an obviously Germanic name ('Loo-temba*ch*er) Frenchified to Lew'tom-bah'shay (approximately). Obviously, one can only give a few

*Jean Charcot (1825–93), French neurologist

examples, but turning now to a famous German name, we have the following, (adopting the same system).

Virchow	'Verchov	'Feershow	'firço:

In general, ignorance of German pronunciation does not lead to serious errors, though there are some curious inconsistencies. Thus, Bach (J. S., the composer) is usually mispronounced Bahk, while Mach (Ernst, the philosopher-physicist whose name has been adopted for the speed scale in relation to the velocity of sound) is definitely Mack (one, two, etc). In fact, the names should rhyme.

In my student days the twisted wires with which surgeons performed craniotomy (the Gigli saw) were referred to as 'giggly' saw. Since the great Italian tenor of the same name became known to all, honourable amends have been made with 'jeel-yee'. This is not the place for a course in the pronunciation of foreign languages and I wish to do no more than draw attention to a deficiency that can give offence or provoke unseemly mirth. A chairman or moderator at an international congress has only, after all, to ask the speaker *sotto voce* how to pronounce his name and then do his level best.

I cannot conclude, however, without brief reference to Turkish. Turkish names do not as yet appear often in medical literature or crop up frequently at meetings. Yet there is one that is forever popping up eponymously because new 'causes' of his syndrome are constantly being discovered. I refer to Behçet. This is usually mispronounced 'baysay', regardless of the fact that in French a cedilla never precedes an e. Be-h-chet is correct with the h sounded separately. The blame for this little difficulty may be laid squarely at the feet of Kemal Atatürk. When he changed the Turkish alphabet from Arabic to Latin characters in 1928, he was most unfortunate in his choice of advisers. The Turkish alphabet is loaded with diacritical marks and the phonetic values bear little relation to those of other languages using Latin characters. So all is forgiven.

As to ensuring that one's name is pronounced correctly after emigrating, even the best-laid plans may go awry. When, in 1823, the 36-year-old Purkyně (pronounced 'Poor-kin-yeh) left his native Bohemia, where the Czech language was his mother-tongue, to take the Chair in Physiology at Breslau, he adopted the spelling Purkinje for his name, to ensure that it was pronounced correctly by the German-speaking population (Breslau had been part of Prussia since 1742). Generations of British and American doctors accordingly pronounced his name Per-'kin-jee. That is the spelling and pronunciation applied to his eponymous fibres in the heart and cells in the cerebellum. Professor J. P. Hill, who taught me embryology at University College, London, did, however, emphasize the correct pronunciation, but without the slightest effect on his pupils. Doubtless the soul of Jan Purkyně is muttering the Czech equivalent of 'You can't win'. But that is not all. After the Second World War, Breslau reverted to Poland; it is spelt Wrocław, and pronounced Vrotswuv. You definitely can't win.

• Foreigners' neologisms

Most of the neologisms perpetrated by foreigners are well-intentioned, if misdirected, faulty extrapolations of grammatical constructions. Rarely the neologisms display a penetrating psychological insight, are more compact than the right word, or fill a gap in the language. The following examples are authentic.

'That man is a scoundler?' A scoundrel, and a swindler; a fine portmanteau word of pure Carrollian quality.

'He is a sneak in the grass.' Worthy of James Joyce?

'I hate my husband passionately!' And why not? Essentially, there is an element of suffering in passion, even if the emotion is love (L. *patior*, *passus*, to suffer; *passionem*).

'We must be precautious.' Improve on that, if you can. The adoption of 'precautious' would enrich the language.
'Buy me an infloatable boat to take to the beach.'

● Ow!

Minerva, writing in the *British Medical Journal* of 23 June 1979, quotes her sister columnist, Ariadne in *New Scientist*, as saying that German pistols go piff-paff, French go pan-pan, Polish pick-puff, and Japanese bakyoom. National differences of utterance extend to other fields. British dogs bark bow-wow and birds say tweet-tweet. French dogs say ouâ-ouâ (wah-wah) and their birds cui-cui (kwee-kwee). German dogs bau-wau (bow-vow). Hungarian birds csip-csip (cheep-cheep). Research carried out between sessions at a congress in Jerusalem disclosed that Israeli dogs bark huv-huv and the birds say tsif-tsif.

Fig. 13 Hieroglyph representing the word 'cat' (pronounced 'mee-oo').

Cats apparently say miaow in all areas. These sensible animals use a universal language and have no need for simultaneous translation at international feline gatherings. Dr John Nunn writes, 'It is, perhaps, even more remarkable that the ancient Egyptian hieroglyphs for cat (Fig. 13) are, to the best of our knowledge, rendered phonetically as mee-oo.[1]

Of greater interest in a medical context are national variations in the vocal response to pain. If pain is sudden and severe, the cry is of animal spontaneity and the same the world over. However, when pain is less severe the vocal response lends itself to a stylized utterance in which there is a component of sympathy seeking. In Britain we say 'ow'. The Germans say 'aua' (owah), the French 'aïe' (eye), the Israelis likewise, the Russians 'oy' and the Hungarians 'jaj' (yoy). Lexicographers, please note!

[1] Nunn, J. (1980). The universal language of cats (correspondence). *Br. Med. J.* **281**, 1430.

• Fingers and digits

My readers would doubtless agree that fingers are part of the human anatomy. Yet there is a certain evasiveness on the part of modern textbooks of anatomy in explicitly stating this fact. Fingers are mentioned, as it were, by implication. Gray's *Anatomy* (33rd edn) states, in the section on osteology, 'The phalanges are fourteen in number, three for each finger and two for the thumb'. Cunningham's *Textbook of Anatomy* and Grant's *Method of Anatomy* exhibit a similar obliquity of approach to the point that we normally have five fingers. I was distressed to find words of advice, in Grant's, that are scarcely appropriate to members of a dignified profession: 'When striking, use the second and third knuckles'. Possibly the writer had inter-hospitals rugby in mind. Lorenz Heister's *Compendium Anatomicum* (1721) is explicit and clearly states, 'We come now to an examination of the hand, which is divided into carpus, metacarpus, fingers, and the little bones called sesamoides, because of their resemblance to grains of sesamum'. Fingers and toes, however, get their due in the *Nomina Anatomica*—namely, the Basel *NA*, 1895; the Paris NA, 1955 (*PNA*); and the Birmingham Revision, 1933 (*BR*).

Fingers are used primarily for grasping and feeling. The word is derived from the same root as fang (the tooth that grasps) and Ger. *fangen*, to grasp. They have secondary uses in counting, measuring, and gesturing. The various names that have been given to the individual fingers in the course of historical times have stemmed in part from these uses. It is of some interest to review these names.

(a) *PNA*: *pollex*, *digitus I*. *BR*: thumb; ultimately from L. *tumor*, swelling; the stout or thick finger; compare Sw. *tumme*, Ger. *Daumen*. The implication of clumsiness in the phrase 'all thumbs' is scarcely merited, since it is the most useful single digit for grasping and the one whose loss causes most disability. The thumb's power is implied in the phrase 'under the thumb of . . . .' Indeed, L. *pollex* derives from L. *pollens*, powerful. Though its use in gesture has not contributed to nomenclature, British custom is the reverse of that in ancient Rome. In Britain thumbs-up signifies 'all well' and thumbs-down 'disaster'. Contrariwise, at the gladiatorial games, thumbs-up from the spectators meant death to the vanquished, while thumbs-down meant 'spare him'. Thumbs-up in Britain was formerly a rude gesture (see middle finger below) and it still is so in much of Europe. Hitch-hikers abroad should beware; thumbing a lift could lead to delays, second only to those of air travel.

(b) *PNA*: *index*, *digitus II*, *BR*: index finger. The forefinger, pointing, indic/ator finger, from its use in the gesture of showing topographical direction. L. *digitus index*, gen. *indicis*; *indicere*, to point out, reveal, inform against—therefore the pointing finger, especially in exposing guilt. Also, L. *digitus salutarius*, the greeting or saluting finger.

(c) *PNA*: *digitus medius*, *digitus III*. *BR*: middle finger. L. *digitus summus*, the tallest finger; *digitus infamis*, disgraceful finger; *digitus impudicus*, shameless, lewd finger (so called because of its use in unbecoming, insulting, or lewd gestures). The 'middle-finger jerk', with the middle finger extended upwards and the remaining fingers flexed, is still widely used in many European countries and in America as an insulting, obscene, or phallic gesture, though in Britain the extended index keeps it company (the Harvey Smith* salute). In Arabic countries the middle finger is extended downwards, and one can only speculate on the reason for this. That the epithet *infamis* was used by Romans without any implication of impropriety is evidenced by the writings of the satirist Persius Flaccus, where he describes how to avoid the influence of the Evil Eye. 'See how granny or aunt, in fear of the god, takes the boy from the cradle and, skilled at averting the evil eye, with her middle finger (*infami digito*) applies the charm with shining spittle on his forehead and little wet lips.'

(d) *PNA*: *digitus anularis*, *digitus IV*. *BR*: ring finger. Rings may be worn on any finger, so why should the fourth finger be so designated? The thumb was favoured in fifteenth- and sixteenth-century Europe; when the Pope names a cardinal, the ring is also placed on the recipient's thumb. In the orthodox Jewish marriage ceremony the ring is placed on the bride's right index. Ecclesiastical rings were also worn on the right index in early times; at a later date it was the left fourth finger, because this was supposed to communicate through a blood vessel with the heart. Yet wedding rings are worn on the right fourth finger in Germany and in those countries, such as Greece, where the Church is Eastern Orthodox. Dextrocardia is not noticeably commoner in these areas. The fourth finger for the wedding ring however, appears to derive from religious practice. 'Weak-man' was used in children's language in my time, but seems to be obsolete. It was so called because the fourth finger cannot be extended while the adjacent fingers are flexed. This restriction arises from fibrous slips connecting its extensor tendon with those of the adjacent fingers.

---

*Harvey Smith was a British champion in show-jumping horsemanship, who achieved notoriety by contemptuously raising two fingers at the judges.

This finger was also named L. *digitus medicinalis*. Isidore of Seville (AD 560–636), the late-Latin lexicographer and etymologist, explains this epithet by the fact that 'with this finger doctors stir pounded ointments'. One can speculate that the choice of this finger for medicinal purposes may have been based on magical or superstitious beliefs regarding the unsuitable potential of the other fingers. Thus thumb was too powerful; the index was for exposing guilt; the middle finger was for obscene use, and the little finger was for scratching one's ear. Thus the fourth finger alone remained for so exalted a purpose as ministering to the sick. Superstitious beliefs die hard. As a recent candidate for a degree course in Classics said, 'My granny told me always to rub in skin cream with that finger rather than the index, because that one was used by Judas to point out Christ'. The same exclusion process may have guided the choice of the fourth finger for ecclesiastical and ultimately marriage usage. Incidentally, a shortened form of *anulus*, Latin for ring, is anus, which survives as the term for the distal intestinal orifice, appropriately adopted from the ring formation of its sphincter.

(e) *PNA*: *digitus minimus*, *digitus V*. *BR*: little finger. Pinkie, mainly Scottish and children's usage (compare Dutch *pink*). L. *digitus minimus*, the smallest finger; *digitus auricularius*; F. (*doigt*) *auriculaire* is normal French usage. The earliest French reference is in *Pantagruel* by Rabelais (1532), and is presumably borrowed from the Latin. The epithet derives from the habit of inserting the little finger tip and its nail into the auditory meatus, a custom which is being superseded by the judicious use of the wire paper-clip (see *Scalpel*; Fig. 15, p. 87).

It is noteworthy that all the Germanic languages have a special word for the digits of the feet (Eng. toe, Ger. *Zehe*, Dutch *teen*, Sw. *te*, and so forth), while Latin and the Romance languages derived therefrom do not (L. *digitus pedis*, It. *dito del piede*, Sp. *dedo del pie*, F. *doigt de pied*).

Movements of the hands and fingers often occur

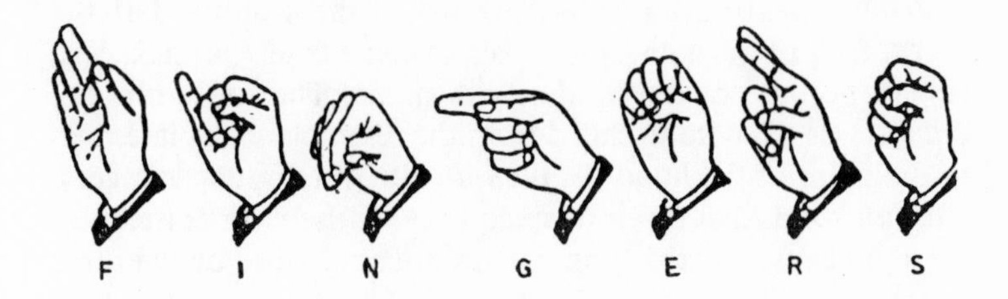

Fig. 14 Deaf-and-dumb alphabetical hand-signals spelling the word 'fingers'.

spontaneously during speech, and in so doing may convey additional meaning or significance to what is said. Formalized movements taking the place of the spoken word are seen in the deaf and dumb alphabet (Fig. 14). The finger's breadth as a clinical measure persists despite attempts to substitute standard units. One medical writer, while deploring the practice, suggested that one should at least use the patient's fingers as units, and not the doctor's. Despite a wholesome trend towards standard units, no one is going to discard 'duodenum' on the ground that it is so called because its length is 12 finger-breadths. Nor would the Germans do so; they call it, even more explicitly *Zwölffingerdarm* (twelve-finger-bowel).

The German for thimble is the same word as the German for foxglove, *Fingerhut* (literally 'finger hat') a plant whose flower closely resembles a thimble in shape, and from whose leaf a glycoside, DIGITALIS, valuable in treating heart disease, is obtainable. Leonhard Fuchs, the German herbalist who gave his name to the fuchsia, noting that the foxglove lacked a Latin or Greek name, proposed in 1542 the name 'digitalis,' from L. *digitabulum*, thimble. But whence comes the 'fox' of foxglove? It is tempting to derive this from the name Fuchs, which is German for fox but, alas, 'foxglove' long antedates this botanist—which just shows the dangers of intuitive etymologizing. The fox is probably related to Norwegian *revbjelde*, fox-bell, and here one can only guess at a mythological association.

While we are botanizing it is worth mentioning that the DATE, fruit of the date palm, takes its name from Gk. *daktulos*, finger, because of the similarity in shape. The PALM tree (L. *palma*) was likewise so called from the resemblance of its leaves to the palm of the hand (L. *palma*) with outstretched fingers. *Daktulos* has also given us dactyl (-⏑⏑), the metrical foot in verse, because the three phalanges are analogues of the three syllables.

The use of fingers for counting has given us the numbers 1 to 10, from which we have the numerals 0 to 9, and the decimal system in calculation and in the recording of scientific measurement. Not surprisingly, these numerals are called digits (fingers). Thus we have DIGITAL readouts on many instruments – including, for example, digital clocks, which are clocks with digits but without hands. Had six fingers been the normal complement of each hand, we would be routinely using a duodecimal system and with it certain arithmetical conveniences. After all, the shilling,* the linear foot, the angular degrees of a circle, the hours of the day, and the minutes of the hour are all divisible exactly by three. And did not Albert Einstein say that God was a mathematician? If so, He might have given us 12 fingers while He was about it.

• Left and right

Many words are shared by the medical profession and the laity. Of these there can be few more fundamentally important than the words LEFT and RIGHT. People sometimes confuse these terms when attempting to describe a position or direction, saying or writing one when they mean the other. An error here can be disastrous, sending the traveller down the wrong road,

* In use when this was first written; equivalent to 12 old pence or 5p.

or causing the surgeon to operate on the wrong side of the body. The annual reports of the medical protection societies give sombre evidence for the latter type of error. Clinically, there seem to be two sources of confusion. Firstly, when we face a patient, their left side is on our right hand. Secondly, and on a more profound level, there is no immediately obvious way of telling which side is which other than by memorization. A minority of people find spontaneous memorization difficult. These people are apt to say 'Turn left' when they mean 'right', or at best will be afflicted with a momentary uncertainty when a traveller asks the way. Some overcome their difficulty by wearing a mnemonic ring on the left hand or adopting an analogous device. It is said that Sigmund Freud would make small writing movements with his right hand to establish his laterality. Yet to the vast majority of people not encumbered with this disability, the identification of right and left by memorization is self-evident and taken for granted.

Well, how *do* we tell left from right? Perhaps the dictionary can offer a definition of these terms. Let us see what the *OED* has to say.

> LEFT adj., the distinctive epithet of the hand normally the weaker of the two and the other parts of the same side of the human body, hence also of what pertains to the corresponding side of any body or object. (A corresponding definition is given for RIGHT.)

As about one person in eleven is left-handed, but only one person in about ten thousand has situs inversus, the *OED* might do better to invoke the cardiac apex or spleen as the left-sided indicator, or for that matter any of the other innumerable constant asymmetries in the structure of animals and plants. However, any such definition implies a knowledge of the structure of living things, and furthermore, this type of lexicographic approach is, in essence, saying that the left side is on the left.

Would the concept of left and right be tenable in a world in which all living things were bilaterally symmetrical, or one in which there were no living things? Surely there must be a pure geometrical or mathematical way of defining laterality! The eighteenth-century German philosopher Immanuel Kant attempted to elaborate the problem of a geometric laterality in space independently of ourselves, and the same problem has vexed philosophers and physicists to this day. Imagine a being out of sight but with whom we can communicate. We do not know if this being is physically symmetrical or, indeed, anything about his anatomy. Before him lies a pair of gloves. Tell him to pick up or otherwise indicate the left-hand glove—if you can. Without an asymmetrical structure which both the questioner and the questioned can observe (such as a constellation), there would appear to be no way of providing an operational definition of left and right. Whatever can exist or occur left-handedly can do so right-handedly. This is one aspect of a concept known to physicists as 'parity'.

Parity 'fell', as the physicists say, when Mrs Chien Shiung Wu carried out a historic experiment in 1952. Radioactive cobalt-60 was cooled to near absolute zero temperature and then submitted to an electromagnetic field, whereupon more electrons were emitted from the south magnetic pole of the nucleus than from its north pole. For the first time it was thus possible to make an absolute labelling of a magnetic pole. With this newly discovered asymmetry it was possible to erect a structure of right and left that could be communicated to our isolated being (unless he was in a galaxy made of anti-matter), and even to a lexicographer, assuming they each had a supply of cobalt-60 and the necessary electrical and low-temperature equipment. In the history of physics there have been some striking reversals of securely held concepts, and he would be a bold man who dares to say that parity has fallen forever. The conundrum of defining right and left may yet return to haunt us.

Lacking as we do the amenity of wearing a miniaturized laboratory on the left wrist, and in general not experiencing undue difficulty in differentiating left from right, we are forced to the conclusion that this innate capacity is ultimately dependent on the asymmetry of our bodies. It is therefore appropriate to turn our attention to the phenomenon that, where asymmetry exists in the structure of the organisms of a species, the asymmetry is constant—that is to say, it is in the same direction, in all members of that species (with only rare exceptions). I refer here to the asymmetries which are observable, for example, in the visceral anatomy of vertebrates; the claws of some crustacea, the spiral shells of some molluscs, and the helical turn of tendrils and some stems. It is not enough to ask why these structures are asymmetrical. We must also ask why the asymmetry is constant in direction. In the course of embryological development there must be a force which turns the enlarging structures in the same direction in all members of the same species. Were this not the case, roughly half the population would have situs inversus, just as a coin will fall heads or tails with equal probability (in the absence of a biasing force).

The work of Afzelius on the immotile cilia syndrome has thrown some light on the area from which this force may emanate, since about 50 per cent of the subjects of this syndrome have situs inversus, often presenting as Kartagener's syndrome. The laevo-amino-acids, which are universal in the chemistry of all living things, build up to subcellular elements whose configuration is helical and of constant laterality. This may be connected with the rotatory component of ciliary action. Ciliary coordination of the viscera in the embryonic archenteron might thus provide the turning force for that constant anatomical asymmetry on which our sense of right and left is based. Retropolating (to coin a phrase) to the origin of life on earth, it was probably by chance that our Primordial Aminoacid progenitor (PA would seem a suitably filial abbreviation) was laevo, though a bias has been suggested. It is a curious thought that had PA been the enantiomorph (mirror-image molecular structure) our hearts would all be right-sided and a reversal of existing asymmetries would hold throughout the animal and plant kingdoms. What we would then have called left and right is best left to professional philosophers!

This is not the place to discuss the cultural sequelae of handedness whence are derived the associations of right with what is adroit, correct, just, proper and straight, and left with what is sinister and inappropriate. These associations have, however, spilled over into the terminology of direction. Some years ago when motoring in France, I enquired the way and was told, '*Tout droit, tout droit*'. I was about to turn right but was saved by my informant adding, '*au fond de la rue*'. We likewise may confuse the foreigner by saying, 'Carry right on,' unless we add 'to the end of the road'. Dutchmen please note; '*Rechtdoor. Tot het eind van de weg*'. No confusion here with '*Gerade-aus*'. Finally, the story must be told of the schoolboy who translated the royal motto '*Dieu et mon droit*' as 'My God, you're right!'

• Disorientation

DISORIENTATED, in a clinical context, means mentally confused; being grossly mistaken regarding one's place in the environment or, by extension, in time. Disorientation may also rarely occur in healthy people during difficult flying conditions.[1,2] Etymologically, disorientated means not knowing in which direction east lies; L. *orire*, *orientem*, to rise; hence the region of the globe beyond that point on the horizon at which the sun rises is the orient, the East (as viewed from

Europe). But why not the West? Why should a confused person not be described as 'disoccidentated' (not knowing in which direction west lies)?

ORIENTAL and OCCIDENTAL date from the fourteenth century. Most European travel during the preceding century was eastward (the crusades, Marco Polo), and thoughts on far-off lands were turned in that direction. Perhaps the clue lies there. Homologues of 'disorientated' exist in all Romance languages (F. *desorienté*, Sp. *desorientado*)—with one curious exception, Portuguese, in which language the equivalent term is '*desnorteado*'. Linguistically, the Portuguese, when confused, do not know in which direction north lies. This probably stems from the days of the great Portuguese navigators of the early sixteenth century, notably Vasco da Gama and Magellan. For direction-finding by night, they would have depended on Polaris, the North Star.

[1] Harding, R. M. and Mills, F. J. (1983). Function of special senses in flight. *Br. Med. J.* **286**, 1728–31.

[2] Braithwaite, M. G. (1985). Disorientation in army helicopter operations: a review. *J. Roy. Soc. Med.* **78**, 856–9.

• Calliper and calibrate

CALLIPER or caliper in orthopaedics is a device applied to a limb, comprising two rods joined together proximally and each turned inward at the distal end to enter an attachment, such as a socket in the heel of a shoe. Also an obsolete instrument for measuring the dimensions of the female pelvis (hence usually called a pelvimeter), similarly comprising two curved arms hinged at one end. The word stems from CALIBRE, originally the diameter of a bullet or cannon ball, and by extension, the internal diameter of a gun, and later of any cylindrical structure, such as an artery or bronchus. Whence calliper, an instrument for measuring the calibre of a gun or other hollow structure, or the external diameter of a convex body and, by analogy, the orthopaedic device defined above. CALIBRATE was originally to measure the calibre of a thermometer tube, whence to graduate a gauge or scale of any kind with allowance for its irregularities.

● -scope

Everyone knows that words ending in -scope (Gk. *skopein*, look at) are concerned with seeing and looking; likewise the related endings -oscopy and -oscopic. The MICROSCOPE (Gk. *micros*, small) and the TELESCOPE (Gk. *tēle*, far off) were both invented about 1610, but the word 'microscope' was not introduced into the English language until 1656 by Thomas Hobbes, the political philosopher. 'Telescope' entered the language a year later; it had been coined in Italian by Galileo Galilei (*telescopio*) and appeared in England in neo-Latin *telescopium*. When the need for a new word arises, several prototypes may be suggested, one of which eventually achieves general adoption. Johann Kepler (1571–1630) suggested the following alternatives for telescope: perspicillum, conspicillum, specillum, and, remarkably, penicillium (L., small pencil) from its shape and dimensions.

In medical instrumentation, -scopes are named according to the part inspected, with a combining o joining the moieties, as in ophthalm-, laryng-, bronch-, aur(i)- or ot-, proct-, sigmoid-, and so on, and collectively as ENDOSCOPES (Gk. *endon*, within). The exception to this is the telescope, as used in rigid, pre-fibre-optic bronchoscopes.

Some -scopes appear to have nothing to do with seeing or looking. Here the metaphorical sense of *skopein* (examine, enquire) is used; for example GYROSCOPE, an instrument designed to examine the dynamics of rotating bodies; hygroscope, an instrument which indicates without actually measuring atmospheric humidity (like the toy man and woman who swing in and out of their cottage doors with changing humidity); it is probably obsolete, but persists adjectivally in HYGROSCOPIC, having the capacity to absorb moisture from the air; and HOROSCOPE (Gk. *hora*, hour) examination of the heavens as they were at the hour of a person's birth.

Of all the non-optical -scopes, the one that many must find puzzling is the STETHOSCOPE (Gk. *stethos*, chest), an instrument for examining the thorax by auscultation. This was Laënnec's* coinage, but here, too, several alternatives were suggested. In his *Traité de l'auscultation médiate* he wrote, 'I have heard it designated by various epithets, all improper, and some barbarous; among the rest by those of sonometer, pectoriloquer, thoraciloquer, medical horn, etc. I have, therefore, given it the name of stethoscope, which appears to me the best to express its principal use'. Another failed contender was stethophone. Of course, the stethoscope may be applied to parts other than the chest to aid recognition of vascular abnormalities in the cranium and elsewhere, and in a negative sense, by the absence of bowel sounds in the abdomen. Indeed, Hildred Carlill, a neurologist at Westminster Hospital in the 1920s and 30s, opined that the main uses of the stethoscope were in emergencies. The flexible tubing could be used as a tourniquet, and the rigid ear-tube for a tracheostomy. In those days senior hospital staff often exhibited showmanship, and the cultivation of minor eccentricities of behaviour or opinion was a relatively harmless way of so doing.

A notional, but none the less valuable, instrument is the RETROSPECTOSCOPE, which enables us to learn from our mistakes. In Amsterdam some years ago I saw a brass plate

*René Laënnec (1781–1826), French physician, invented the stethoscope.

on a door bearing the occupant's name and occupation—PSYCHOSCOPIST. Now, if he had a psychoscope, that was the very thing I needed. I rang the bell. There was no response, so I shall never know. Well, anyway, it was probably just a crystal ball.

● Forceps

FORCEPS is derived from the Latin *formuceps*, (*formus*, hot; *capere*, seize), originally an implement for seizing hot objects. It was used in this sense for retrieving hot chestnuts from a coal fire, until the Clean Air Act put a stop to coal fires; likewise for removing instruments from boiling water in a sterilizer, another practice that has virtually ended in Britain's hospitals by the introduction of central sterile supplies. The implication of the seizure of a hot object is long since obsolete. In the surgical context, a forceps is an instrument comprising two opposing limbs for seizing, and then holding, pulling, compressing, or nipping off. The word may be used as a singular or a plural. In Latin it is singular; the plural is *forcipes* (three syllables). The terminal s of forceps has been mistakenly adopted as indicating plurality, possibly by association with scissors and pliers.

Considered mechanically, there are three basic types of forceps; (a) pivoted at a point between the extremeties (for example, dental, obstetric, artery); (b) sprung at the proximal end (dissecting, swab-holding, splinter, 'tweezers' and (c) cross-action, where the jaws open when the shanks are approximated, and close when the pressure is released (towel clips, Dieffenbach's). The seizure of a tissue by forceps is obtained initially by manual application, but may be maintained by a self-locking catch or ratchet. Where a locked-on forceps is applied with great pressure, as for example with artery forceps, the instrument is better called a clamp. Additionally, when not used to grasp a tissue, a forceps is called a 'holder', as in needle holder.

The sales catalogue of Downs Surgical plc lists 141 varieties of forceps. Nomenclature falls into three main categories; (a) functional (for example, artery, dissecting, tissue); eponymous (Spencer-Wells', Moynihan's, Kocher's); and (c) zoomorphic (mouse-tooth, bulldog, alligator, mosquito). All the examples are adjectival, and followed by 'forceps'. I recall Sir Wilfred Trotter handing back toothed forceps with the words, 'Not these, Sister. The edentulous ones, please'.

Now for a little etymologizing on some terms mentioned. TWEEZERS, from 'etweeze', an anglicization of the French plural of *étui*, a case or box, especially one for small articles such as needles, bodkins, toothpicks; later, a case for surgical instruments, and finally the instrument itself. ALLIGATOR, from the Spanish *el lagarto*, the lizard; ultimately from the Latin, *ille lacertus*. MOSQUITO, the vector of various viral diseases and of malaria and filariasis; unchanged from the Spanish (and interestingly, retaining the Spanish silent u); a little fly, diminutive of the Spanish *mosca* (L. *musca*). It remains a mystery how or when the forceps acquired the epithet mosquito, but one can guess that it was based on the capacity of both to give a little bite.

• Trocar and cannula

TROCAR and CANNULA are a closely associated pair; they are almost uttered as one word. A trocar is a perforator whose function is to facilitate the insertion of a cannula. It lies snugly in the cannula's lumen, the sharp end projecting beyond the cannula's rim. In transverse section the sharp end is an equilateral triangle whose angles form the cutting edges and whose sides form the flats. 'Trocar' derives from the French *trois carres*, meaning three sides, and by implication, three edges. *Carre* is the breadth of a square, but may be extended to the sides of other polygons and the angles between them, for example, *lame à trois carres*, three edged sword. The English adjective square and the French *carré* have a common origin in the Latin *ex quadra*.

A cannula is a rigid tube for insertion into a body cavity or vessel. The word is derived from L. *canna*, reed. The reeds are members of the grass family (*Gramineae*), and as such have stems which are hollow between the nodes. This was common knowledge even to botanical ignoramuses in the days when drinking straws were made of straw. Just as cannon (the sort that fires projectiles) is etymologically the augmentative of *canna*, so 'cannula' is its diminutive. It is the rigidity and hollowness of the reed that justifies the term 'cannula' as a little reed. Somewhere between the grasses in size and the bamboos (all *Gramineae* they) lie the canes, the nearest word we have in the vernacular to *canna*, whence we have (a) cane sugar (sucrose), which is bad for the teeth and also, we are assured, the coronary arteries, and (b) the cane, used as a light stick, and carried as a social gesture or wielded punitively on schoolboy rumps.

• Bougie

Everyone in this country who has learnt a foreign European language knows that there are words that have a common origin and may have an identical spelling, but which have different meanings. For example *Smoking* in German means dinner jacket (Amer. tuxedo), and *les girls* in French are not the sort that join the Guides over here. Now, the French *chandelle* is a large ornamental candle, and, as every schoolchild knows, it is the sort that *le jeu ne vaut pas*. The currently used French word for candle, *bougie*, arose in the following way. The Algerian port Bejaia was for long the centre of the wax trade, and candles were a major item of export. Even before the French took control of Algeria in 1830 they called Bejaia 'Bougie' and the wax candles they imported therefrom came to be called *bougies* after their place of origin. Because of their pliability at body temperature, slender candles were used by surgeons to dilate strictures of the urethra and of other passages and orifices, being occasionally 'medicated' with caustic at the tip. It was required of these candles (*bougies*) that they be smooth and supple, yet sufficiently firm not to bend or twist under pressure of insertion.[1] The right consistency was obtained by melting 2 parts beeswax, 6 parts olive oil, and 3 parts minium (red lead). Stiffening could also be imparted by the additon of litharge plus a trace of cinnabar, or by increasing the proportion of beeswax. The wick cannot have ensured freedom from the risk of breakage *in situ*, and *bougies* were therefore reinforced with linen, strips of which were dipped in the melted wax, then folded and rolled on a marble slab until smoothly cylindrical.[1] Rigid metallic *bougies* are preferably called dilators,[2] though the function of a pliable *bougie* is also that of a dilator; nevertheless, in surgical practice the distinction between the pliable and the rigid seems to be important for nomenclature. It remains to be said that, if you wish to buy a candle in a French-speaking country, ask for a *bougie*.

[1] Morris, R., Kendrick, J., and others (1807). *The Edinburgh medical and physical dictionary*. Bell & Bradfute; Mundell, Doig & Stevenson, Edinburgh.

[2] Kirkup, J. (1985). The history and evolution of surgical instruments. **IV** Probes and their allies. *Ann. Roy. Coll. Surg. of England*; **67**, 56–60.

• Scissors

There is nothing especially noteworthy about the derivation of 'scissors', but as you have read this far, we may as well continue. The word stems from L. *cisoria*, plural of *cisorium*, a cutting instrument; the plural usage presumably stems from the two blades. The ultimate root is *caesum*, past participle of *caedere*, to cut. But what interests me is the plural usage for single objects having a paired structure, and for which there is no current singular. This is manifest in the names of nether garments (trousers, pants, drawers, knickers, bloomers, jeans), spectacles, and instruments (scissors, pliers, tweezers, shears). There is no such thing as a pant, a trouser, or a scissor in the same sense as the plural form. Singular forms do exist for some of these terms, but they have other meanings.

• Scalpel

The small surgical knife, known as SCALPEL, has been in use at least since Graeco-Roman times. The word is almost unchanged from the Latin *scalpellum*, the suffix *-ellum* indicating the diminutive of *scalprum*, a sharp cutting instrument, as used by farmers, shoemakers, and sculptors (evidently a larger, cruder implement). The verb *scalpo*, to cut, scrape, engrave, merges with *sculpo*, which has much the same meaning but includes 'to form by carving', whence *sculptor*, a word that has entered the English language unchanged.

Sadly, two useful words have been lost on the way, *scalptorium* and *auriscalpium*. The *scalptorium* was an instrument in the form of a hand for scratching oneself. I am the grateful recipient of one of these back-scratchers; it is made of an African hardwood and its distal extremity is shaped into

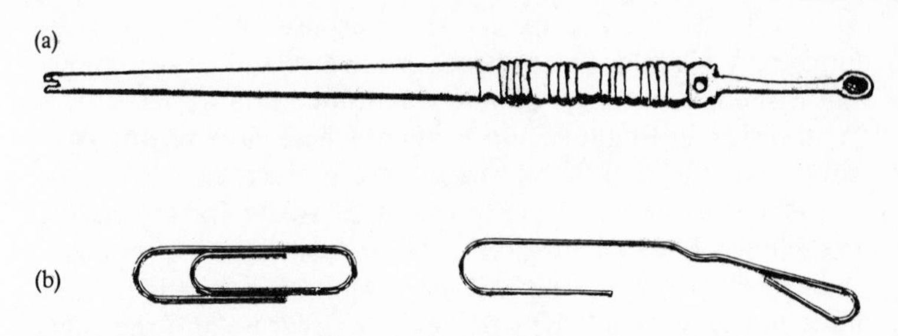

Fig. 15 Auriscalpium: (a) bronze Roman ear-pick (British Museum); (b) wire paper-clip unfolded for use as an ear-pick.

a small hand whose partially flexed fingers are nicely adapted to reach that not-too-accessible interscapular area. The *auriscalpium*, another toiletry article, was an ear-pick, so useful for extracting cerumen and for obtaining relief from itching (Fig. 15). The handy modern alternative for personal use seems to be a modified wire paper-clip.

• Plaster of Paris

As distinct from naturally occurring gypsum (Gk. *gypsos*; $CaSO_4.2H_2O$), gypsum plaster (PLASTER OF PARIS) is principally the hemihydrate ($CaSO_4.\ \frac{1}{2}H_2O$), a product obtained by heating gypsum to drive off water of crystallization. When water is mixed with plaster of Paris, it forms a solid mass of interlocking crystals and is reconstituted as the fully hydrated form, with which every doctor is familiar. This phenomenon was known in antiquity to the Egyptians and Greeks, who used it for building and for lining walls. Splinting of limb fractures by means of plaster of Paris was practised by Arabs as early as the tenth century, but was not effectively used in Europe for this purpose until the plaster-impregnated bandage was introduced in Holland by Anthonius Mathijsen in the 1850s.[1] There was some initial resistance to its adoption in England and France, where slow-setting egg-white and starch on linen strips were preferred.

In English the word gypsum is used solely for the native mineral; likewise, *le gypse* in French. By contrast, in most other European languages there is no terminological distinction between the mineral, the calcined hemihydrate and the final product as applied to architectural, artistic or surgical use, where the term used is a derivative of the Latin *gypsum*. Thus it is *gips* in Dutch, German, the Scandinavian languages, and Russian; *gipsz* in Hungarian, *kips* in Finnish, *gypsos* in modern and ancient Greek, *gesso* in Italian and Portuguese, and *yeso* in Spanish. The French call the secondary products *plâtre de moulage* (plaster for moulds), but in surgical practice usually just *le plâtre*.

Only in English are the hemihydrate and final product called plaster of Paris. This stems from the large deposits of gypsum in and around Paris, at Montmartre in particular, where it has been mined since the twelfth century. France is still the largest producer of gypsum in Europe. The Parisian epithet was probably introduced into English by Henry III who is said to have imported plaster for decorative purposes in 1254 after a visit to Paris. The *OED* quotes,

> (1387) Bysides Parys is greet plente of a manere stoon þar hatte gypsus and is i-cleped white plaistre.

> (1462) The chambyr he lett make fast
> Wyth plaster of parys þat will last.

And even though we now get it from Nottinghamshire, plaster of Paris it still is.

[1] Mathijsen, A. (1852) *Nieuwe wijse van aanwending van het gips-verband bij beenbreuken*. van Loghem, Haarlem.

Glue

The original GLUE (*gluten* in Latin) is now an almost obsolete adhesive made from animal hides, bones, and fish residues. It had a number of disadvantages in practice, being slow to melt for application and, after setting, vulnerable to heat and moisture. It has been superseded by a wide range of synthetic substances which are free from these disadvantages; they are generally known as adhesives. Unfortunately, the volatile organic solvents in which they are applied have tempted people, mainly youngsters, to inhale the solvents to induce a state of quasi-inebriation, a dangerous practice which has led to some deaths. However, the British penchant for monosyllables has brought back the obsolescent word 'glue', since the practice is known as GLUE SNIFFING. We also have GLUE EAR for chronic seromucous (secretory/catarrhal) otitis media.

When the starch is washed out of wheat flour, the residue is a sticky mass which is named GLUTEN, the Latin for glue. Indeed, starch paste is still a common adhesive. Gluten provides the elasticity which imparts coherence to the rising dough in bread-making. Likewise, the specially high gluten content of the wheat variety *Triticum durum* induces the characteristic plasticity of pasta. Unfortunately, gluten induces coeliac disease in those who are endowed with that predisposition. GLUTAMIC ACID, an amino acid widely distributed in tissues (especially in the brain) and indirectly a neurotransmitter, is so called because of its high concentration in gluten. Monosodium GLUTAMATE adds that savoury taste which enables the gastronomic plebs to swallow with relish what would otherwise be tasteless.

The Greeks also had a word for glue, *kolla*. In its English language manifestations the k is changed to c. COLLAGEN, the fibrous protein that holds the body together, is so called from the glue to which it is converted on boiling (*kolla*, with *genēs*, to become). COLLOID was so called from the gummy consistency of the early recognized organic colloids, though the term has now been widened to comprise also a range of non-sticky varieties, and the present definition is based on molecular structure. COLLODION, another gluey substance, is a solution of gun-cotton (cellulose nitrate) in ether and alcohol. Applied in the manner of a paint, it is used to fix small dressings and splints for digital injuries. COLLAGE is a picture made up from scraps of paper and various odds and ends glued to the canvas. It is not an old lady's choker.

No research project may be launched without a statement of PROTOCOL. Originally a protocol was a fly-leaf glued to the case of a volume, and containing an account of its contents (Gk. *prōtos*, first; *kolla*, glue). Over the past four centuries it has left its gluey origin and undergone several shifts and diversifications of meaning, travelling via the original note of a transaction, then the original draft of a diplomatic document (especially one of the propositions agreed to and signed by the parties to be embodied in a formal treaty), to the current medical usage as a precise and detailed statement of the plan for a research project.

• Caucasian

Introduction

The term Caucasian, used to denote white Europeans, is creeping into British medical literature. Since most members of that group stem geographically from nowhere near the Caucasus mountains, the use of the term must perplex many, and irritate quite a few. The term has been largely abandoned by anthropologists,[1] but has been used until recently by immigration and prison officials in the United States of America; it is still in regular usage in American medical literature.

In any clinical or epidemiological context it may be as important to record a person's ethnic origin as, for example, his occupation. Fourteen per cent of the population of the USA is now non-white[2] and some form of racial designation is often necessary. In British medical writing until the mid-twentieth century, a person was assumed to be Caucasian unless otherwise stated. Since then the influx to the United Kingdom of persons of African, Indian, Chinese, and other antecedents has now raised the proportion of non-whites to 5 per cent.[3] Hence the intrusion of 'Caucasian' into our medical literature. Is this a suitable epithet and, if not, what are the alternatives? And anyway, why 'Caucasian'? What follows has been conceived in a medical context; it may not tally with current thought in anthropology or sociology.

Origin of the term

Geographical exploration by Europeans in the eighteenth century led to recognition of the physical differences that exist between widely separated populations. Various classifications of these differences were proposed, and that of Blumenbach was widely accepted. Johann Friedrich Blumenbach (1752–1840; Fig. 16) professor of Medicine at Göttingen, was one of the founding fathers of physical anthropology. He recognized that plants and animals were capable of becoming modified in form as a result of environmental changes. He believed that variations (*degeneratio*) were derived from a primary variety (*varietas primigenia*)[4], and that the races of mankind had thus been derived from the 'white European' variety which he called *Caucasiana*.

He published his classification of the races of mankind in *De generis humani varietate nativa liber* in 1776, and it was in the third edition (1795) of this work that he described the five varieties of races as Caucasian, Mongolian, Ethiopian (that is, African), American and Malay.[5] The English translation by Bendyshe gives his reason as follows; 'I have taken the name

Fig. 16 Johann Friederich Blumenbach (1752–1840), physician and anthropologist. (Courtesy of the Wellcome Institute Library, London.)

of this variety from Mount Caucasus, both because its neighbourhood, and especially its southern slope, produces the most beautiful race of men, I mean the Georgian; and because all physiological reasons converge to this, that in that region, if anywhere, it seems we ought with the greatest probability to place the autochthones of mankind. For in the first place,

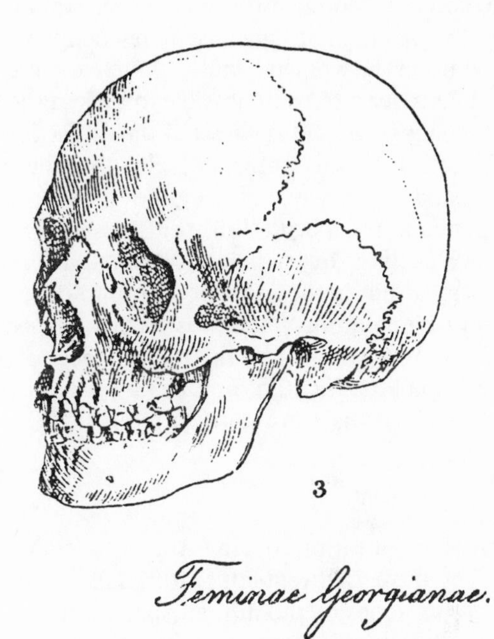

Fig. 17 The Georgian female skull that Blumenbach admired, and that contributed to his naming the 'white European' race 'Caucasian'. (From Plate 14 in his *De generis humani varietate*, 1776.)

that stock displays, as we have seen the most beautiful form of the skull, from which, as from a mean and primeval type, the others diverge by most easy gradations on both sides to the two most ultimate extremes (that is, on the one side the Mongolian, on the other the Ethiopian). Besides, it is white in colour, which we may fairly assume to have been the primitive colour of mankind . . . '[6] (Fig. 17).

Blumenbach travelled little but was obviously influenced by those of his contemporaries who had. He quotes Chardin thus; 'The blood of Georgia is the best . . . perhaps in the world. I have not observed a single ugly face in that country . . . Nature has lavished upon the women beauties which are not to be seen elsewhere. I consider it to be impossible to look at them without loving them'. It is on the flimsy basis of the subjective responses of these eighteenth-century anthropologists that the ethnic term Caucasian stands. It is easy, in the light of present knowledge, to smile at such naïve speculations, but these must be seen in historical perspective. Nevertheless, Blumenbach made many valuable observations in the fields of anthropology, physiology and comparative anatomy, and was rightly respected in his time. This may account for the persistence of the term Caucasian despite the fact that no one now seriously suggests that all white Europeans are derived from the Georgians of the Caucasus.

Taxonomy

Geographical exploration during the late eighteenth, the nineteenth and the early twentieth centuries led to the discovery of ever more varieties of the human species, and close study of these and of previously known races with respect to skeletal measurement, skin pigmentation, shape of nose, hair coil and other anthropometric features in turn led to complex and detailed classifications.[7–9] A geographical gradient of certain characters (cline) between adjacent populations may render any rigid taxonomical subdivisions merely arbitrary. Chemical and molecular analysis of proteins has shown that racial classifications do not match well with most patterns of gene frequencies in human populations. Modern physical anthropology has greatly reduced its earlier emphasis on racial classification based on external appearance and general morphology. In 'How different are human races?', J. S. Jones, of University College, London, writes, 'The geographical trends of gene frequency for a sample of human polymorphisms hardly ever parallel those for skin colour or body form . . .

Only about 10 per cent of the total biological diversity of mankind arises from genetic divergence between "racial" groups'.[10]

Race and disease

Where do we stand in the context of the clinical and epidemiological need for the determination of race? It is well known that there are geographical/racial variations in the prevalence of certain genetic biochemical differences which are of clinical importance. This is manifested in certain diseases, and in the response to drugs in the treatment of diseases which may not themselves be genetically determined. Some examples follow. The abnormal haemoglobin, which is due to the sickle cell gene, is probably responsible for more adult illness and infant deaths than any other inherited disease. It is found in up to about a quarter of the Negro population in an east-to-west belt across central Africa,[11] and in many descendants of African origin living elsewhere. Haemoglobin C causes a similar but less severe illness, mainly in Burkina Faso (formerly Upper Volta) and Ghana;[12] haemoglobin E does likewise in Thailand. Thalassaemia affects children in a geographical band across the Mediterranean, Middle East and Indo-China.[13] Adult intestinal hypolactasia is present in over 90 per cent of Japanese, Chinese, Thais, Amerinds, and some African tribes, whereas in north-west Europe it is present in 12 per cent of the population, and in less than 4 per cent of Scandinavians.[14] Phenylketonuria is rare in Negroes and Japanese, and virtually absent in Ashkenazi Jews, yet fairly common in oriental Jews.[15] Tay–Sachs disease and essential pentosuria are almost confined to Ashkenazi Jews.[15] Knowledge of a patient's ethnic origin may also be important in the treatment of disease. Glucose-6-phosphate dehydrogenase deficiency in the red cells, which is prevalent in Greece, Sardinia, the Middle East and India,[16] may be complicated by severe haemolysis following ingestion of certain drugs (various anti-malarials, sulphonamides.

sulphones, and nitrofurans.[17] The rate of inactivation of drugs by acetylation depends on whether the patient is a 'fast' or 'slow' acetylator.[18] Slow acetylators are at greater risk of side-effects when taking isoniazid, phenelzine, sulphonamides, hydrallazine or dapsone.[19] Slow acetylator status predominates in Egyptians[20] and Ethiopians (83 per cent),[21] is present in about 50 per cent of Europeans, but is uncommon in Japanese and Amerinds and rare in Canadian Eskimos.[22] Enough has been said to show the desirability of establishing a patient's ethnic origin both for diagnosis and treatment. This not only applies to conditions in which a connection is recognized, but also doubtless, to as yet undiscovered disorders and variations in response to drugs. 'Consideration of ethnically determined differences in drug metabolism highlights the potential dangers of extrapolating research data generated in one racial group and applying it to another.'[23]

Naming the races

What designations should be used in clinical and epidemiological contexts? In epidemiological studies there may be time for elaborate methods of identifying gene frequencies, but in the hurly-burly of clinical practice there is at present no room for these, and we may be forced to depend on the five (or so) traditional nineteenth-century racial groups.

The naming of races is bedevilled by emotional overtones. Pejorative implications in a designation lead to a change or series of changes in that designation. As Jules Feiffer, the cartoonist, has put it, 'As a matter of racial pride we want to be called "blacks", which has replaced Afro-American, which has replaced negroes, which has replaced coloured people, which has replaced darkies, which has replaced blacks'.[24] What is needed are names that cannot easily be contaminated by derogatory implication, but where the meaning is clear and self-evident. This is best done on a geographical basis. Accordingly, Blumenbach's choice of Caucasian for the 'white European' race is inappropriate, and should be abandoned.

It might be argued that, as all concerned know what Caucasian means, a change is unnecessary and that a name does not need to be descriptive of origin or condition. After all, no one now believes that influenza is caused by astral influence, or that malaria is due to bad air. This argument does not hold for Caucasian. Because race is primarily based on geographical location, racial designations should not be such as to risk confusion with unrelated geographical names. D. M. Lang writes, 'In view of widespread misconceptions, a word must be said about the term "Caucasian" itself. Certain physical anthropologists who should know better, and also American immigration authorities who cannot be expected to do so, habitually use this word to denote anyone who is not a Negro, Indian, Chinese—in fact, as the virtual opposite of "coloured" . . . . This is utterly unscientific, and a complete misnomer. The Anglo-Saxons, Latins, Slavs, and others to whom the term is so loosely applied have absolutely no historical or ethnic connexion with the Caucasian peoples proper'.[25]

Let us see what our European colleagues say about 'Caucasian'. The *Grand Larousse* says, *se disait autrefois* (was formerly used). *Trésor de la Langue Française* says, *vieilli* (obsolescent). *Dizionario Enciclopedico Italiano* says, *termine usato talora, sopratutto in passato* (term used sometimes, especially in the past). *Grande Dizionario della Lingua Italiana* says, *oggi non più usato dagli scienziati* (no longer used nowadays by scientists). A Swiss colleague said he had seen the term so used in French and German papers in recent years, but solely by authors who had been extensively exposed to American literature. There can be no serious doubt that the use of Caucasian in British medical publications has been imported from America. I imagine that Blumenbach's term was adopted in the USA because it was thought to be 'scientific' and therefore less amenable to ideological or emotional manipulation.

If we are to reject Caucasian as a term to denote the 'white European' race, as I think we should, what are the alternatives? By what criteria should a term be assessed? It seems to me that any term used to designate a racial stock should be characterized by the following: (a) a geographical basis derived from the place of origin, (b) an absence of alternative meanings, (c) a single word, and (d) a self-evident meaning. Let us now look critically at some terms in current use, in the light of these criteria.

Caucasian—geographically wrong except for a small minority comprising Georgians, Circassians, Kabardians, Abkhazians, Avars, Lezghians; and others inhabiting the Caucasus.[25]
European geographical race[26]—explicit but unwieldy.
European[27]—excludes those living in other continents;
Caucasoid[7, 26, 28] (*en suite* with negroid, mongoloid, australoid, etc.)—much favoured by American anthropologists. It retains the fallacious Caucasian implication; the suffix -oid (Gk. *eidos*; form, shape, resemblance) also implies a resemblance to some definitive concept, as in mastoid, thyroid, and it is inapplicable here.
White—much used in English-speaking countries. There are, however, varying degrees of skin pigmentation in Caucasians; (compare the loose usage of white respecting coffee and wine).
Europid (*en suite* with negrid, mongolid, australid, etc.)—adopted by Baker[9] following its introduction by Peters.[29] The suffix -id is stated to be a truncation of the Greek -ides, of the family of.
'Europid', which will be unfamiliar to most readers, does fulfil the above-mentioned criteria. Its use in a medical journal might initially evoke more letters of complaint than the use of Caucasian does now, but I believe that, with repeated usage under authoritative aegis, familiarity would achieve acceptance.

Whatever term is adopted, it is unlikely to receive immediate and widespread approval. Analogous dilemmas will confront the keepers of editorial style books with respect to Asian,

Mongoloid and so forth. Persons of racially-mixed origin present further terminological problems. These difficulties will have to be faced. Authoritative editorial persuasion and influence should provide solutions. A conference of influential medical editors might achieve a consensus and impose an acceptable style in this area.

[1] Littlefield, A., Lieberman, L., and Reynolds, L. T. (1982). Redefining race: the potential demise of a concept in physical anthropology. *Curr. Anthropol.*, **23**, 641–55.

[2] Reuter (1983). Blacks to stir democrats. *The Times*, 14 Mar, p. 6 (col. 8).

[3] Office of Population Censuses Surveys. *Labour force survey 1981*, table 5.7. HMSO, London.

[4] Baker, J. R. (1974). *Race*, p. 26. Oxford University Press.

[5] Blumenbach, J. F. (1795). *De generis humani varietate nativa* (3rd edn). Vandenhoek and Ruprecht, Göttingen.

[6] Bendyshe, T. (translator) (1865). *The anthropological treatises of Johann Friedrich Blumenbach*. Longmans, London.

[7] Cole, S. (1963). *Races of mankind*. British Museum (Natural History), London.

[8] Coon, C. S. and Hunt, E. E. (1966). *The living races of man*. Jonathan Cape, London.

[9] Baker, J. R. (1974). *Race*, pp. 624–5. Oxford University Press.

[10] Jones, J. S. (1981). How different are human races? *Nature, Lond.*, **293**, 188–90.

[11] Weiner, J. S. (1977). Human ecology: disease. In *Human biology*, 2nd edn (eds. G. A. Harrison, J. S. Weiner, N. A. Barnicott, and V. A. Reynolds) p. 462. Oxford University Press.

[12] Barnicot, N. A. (1977). Biological variation in modern populations: biochemical variation. In *Human biology*, 2nd edn (eds. G. A. Harrison, J. S. Weiner, N. A. Barnicot, and V. A. Reynolds) p. 230. Oxford University Press.

[13] Weiner, J. S. (1977). Human ecology: disease. In *Human biology*, 2nd edn (eds. G. A. Harrison, J. S. Weiner, N. A. Barnicot, and V. A. Reynolds) p. 463. Oxford University Press.

[14] Simoons, F. J. (1978). The geographic hypothesis and lactose malabsorption. *Am. J. Digestive Dis.*, **23**, 963–97.

[15] Adam, A. (1973). Genetic diseases among Jews. *Isr. J. Med. Sci.*, **9**, 1383–92.

[16] Barnicot, N. A. In *Human biology*, 2nd edn (eds. G. A. Harrison, J. S. Weiner, N. A. Barnicot, and V. A. Reynolds) p. 235. Oxford University Press.
[17] Smith, S. E. and Rawlins, R. D. (1973). *Variability in drug response*, p. 127. Butterworth, London.
[18] Evans, D. A. P., Manley, K. A., and McKusick, V. A. (1960). Genetic control of isoniazid metabolism in man. *Br. Med. J.*, **ii**, 485–91.
[19] Smith, S. E. and Rawlins, R. D. (1973). *Variability in drug response*, pp. 74–6. Butterworth, London.
[20] Hashem, N., Kahalifa, A., and Nour, A. (1969). The frequency of isoniazid acetylase enzyme deficiency among Egyptians. *Am. J. Phys. Anthropol.*, **31**, 97–101.
[21] Russell, S. L. and Russell, D. W. (1973). Isoniazid acetylator phenotyping of Amharas in Ethiopia. *African J. Med. Sci.*, **4**, 1–5.
[22] Armstrong, A. R. and Peart, H. E. (1960). A comparison between the behaviour of Eskimos and non-Eskimos to the administration of isoniazid. *Am. Rev. Respir. Dis.*, **81**, 588–94.
[23] Whitford, G. M. (1978). Acetylator phenotype in relation to monoamine oxidase inhibitor antidepressant therapy. *Int. Pharmacopysychiatry*, **13**, 126–32.
[24] Heller, S., ed. (1982). *Jules Feiffer's America*, p. 108. Penguin Books, Harmondsworth.
[25] Lang, D. M. (1966). *The Georgians*. Thames & Hudson, London.
[26] Garn, S. M. (1963). Anthropology. In *Encyclopaedia Britannica*, Vol. 6, v, 96. William Benton, London, Chicago, Geneva, Sydney, Toronto.
[27] Challacombe, P. N., Wheeler, E. E., Phillips, M. J., and Eden, O. B. (1983). Leishman-Donovan bodies in the duodenal mucosa of a child with kala-azar. *Br. Med. J.*, **287**, 789.
[28] Montague, A. (1977). Anthropology. In *Encyclopedia Americana*: 85. William Benton, London, Chicago, Geneva, Sydney, Toronto.
[29] Peters, H. B. (1937). Die wissenschaftlichen Namen der menschlichen Körperformgruppen. *Z. Rassenkunde*, **6**, 211–41.

• Porter

Everyone knows what a PORTER is, and every doctor presumably knows what a hospital porter is. But did you know that there are two sorts of porter, not merely etymologically, but also in practice? There is the gate-keeper porter (L. *porta*, a gate or door), and the load-bearing porter (L. *portare*, to carry). It is by pure coincidence that these two unrelated roots have joined to give us the same final derived English word. Such is the power of words that, in hospital administration, this unification of terms has enforced a unitary staff structure. Both types of porter come under the single authority of the Portering Manager; thus the distinction becomes blurred. Yet the functional distinction remains. The duties of the gate porter comprise custody of the keys, dealing with enquiries, directing members of the public, maintaining a patient register and ordering hire-cars for staff and patients. He is sited at or near the public entrance door. In a lower grade (except for X-ray and pharmacy porters) are the load bearers or pushers. They convey patients by wheelchair, stretcher or bed, including those on their final journey to the mortuary; they also deliver laundry, stores, meals, furniture, etc.

Whilst it is clear beyond doubt that the porter in a railway station (if he is not by now an extinct species) is a load bearer, the porter in a block of flats is, by contrast, in an ambiguous position. He usually regards himself as a gate-keeper and a cut above the load bearer, and much embarrassment could be avoided if the distinction were clarified to visitors. 'Oh no, madam, I don't carry luggage!' In American English, he is called a janitor (L. from L. *janua*, entrance door) in an apartment block, but despite the terminological distinction, he is likely to include load-bearing among his duties. Hospital administrators will not be troubled by linguistic hair-splitting. Yet at King's College Hospital, London, the 'problem' recently ceased to exist; the gate or front-hall porters were removed from their location and were replaced by lady receptionists.

• Meniscus

Those of my generation whose first camera was a box Brownie will know that it had a single MENISCUS lens. A meniscus camera lens has concavo-convex surfaces and it is thicker at the centre than at the periphery. In section it resembles a crescent moon, whence its name (Gk. *mēniskos*, diminutive of *mēnē*, moon). As students we meet with another meniscus when measuring liquid volumes, and we learn to take into account the curved upper surface of a column of mercury (convex) or aqueous fluid (concave) in a burette. Finally, we encounter the intra-articular meniscus of fibro-cartilage, most clearly seen as such in the knee joint where it is notoriously prone to injury, but present also in the sterno-clavicular, temporo-mandibular, and wrist joints. By contrast with the optical meniscus, the intra-articular meniscus is thicker at the periphery than at the centre; its name derives from its crescentic outline in the knee-joint; hence it is also called the SEMILUNAR cartilage.

Any curved structure like the arc of a circle may be called a CRESCENT; for example, Park Crescent in London, home of the Medical Research Council. This derives from its resemblance to the crescent moon. But why is the crescent moon so called? Because it is the shape of a growing moon between 'new' moon and first quarter, (L. *cresco*, to grow; *crescens*, *crescent-*, growing).

• Nubes

Nubes is Latin for cloud. OBNUBILATION means overclouding and, in a medical context, clouding of the mind. It is a fine piece of Victorian orotund magniloquence, to be uttered sententiously and with *gravitas*. I can see it now. 'Doctor, my husband is very drowsy and talks muddled.' 'Yes, Madam, I fear there is a modicum of obnubilation.' That generation of doctors, well-grounded in school Latin, would have found nothing strange in such terminology.

We still use the corresponding Greek root, *typhos*, cloud, mist, in TYPHUS and TYPHOID, so called because of the clouding of consciousness that is often a feature of these diseases. Another Latin word for cloud is NEBULA, a term used for clouding of the cornea. A device for converting a liquid into an aerosol, in fact, a mist maker, is a NEBULIZER. The corresponding Greek is *nephelē*, whence we have NEPHELOMETER, an instrument for estimating the number of particles in a suspension by measuring the light scattering or cloudiness. It seems odd that the German language, which tends to eschew the use of ancient classical roots, has *Nebel* for mist or fog.

NUBILE does not mean, as many people seem to think, sexually attractive (in the feminine gender); it means of females, marriageable; from L. *nubere* (p.p. *nuptum*), to be veiled in the presence of the bridegroom, hence, to marry; the veil is a garment whose optical properties resemble those of a mist. Hence also NUPTIALS. *His* head may also be in a cloud, but pleasantly obnubilated.

• Shellfish

Shellfish, as carriers of infection, are in the news from time to time. Oysters, cockles, and mussels readily become contaminated when grown in river estuaries polluted by sewage. Oysters, traditionally eaten raw, are freed from bacteria by depuration, but may retain pathogenic viruses. Cockles and mussels are cooked, but may be inadequately thawed, and likewise present a health hazard. The English language, in the richness of its vocabulary, often provides a choice of synonyms. How odd, therefore, that we have in the vernacular one word, namely SHELLFISH, for the two completely unrelated classes, molluscs and crustacea. Smart restaurants found 'shellfish' too plebeian and coined the word 'seafood'; this has spread to the menu of the 'caff', and even to medical literature,[1,2] where it has appeared within apologetic quote marks. 'Seafood' is an even less specific term than 'shellfish', and could be held to include vertebrate fish and aquatic mammals, which it does not. There is a special risk in ordering *Schellfisch* (pronounced shellfish) in German-speaking countries. If you do this, you will get haddock. If you want shellfish or seafood, you should ask for *Schaltiere* or *Meeresfrüchte*.

[1] Public Health Laboratory Service Communicable Disease Surveillance Centre and the Food Hygiene and Virus Reference Laboratores. Illness associated with fish and shellfish in England and Wales, 1981–2 (1983). *Br. Med. J.*, **287**, 1284–5.

[2] Heller, D., Gill, O. N., Raynham, E., Kirkland, T., Zadick, P. M., and Stanwell-Smith, R. (1986). An outbreak of gastrointestinal illness associated with consumption of raw depurated oysters. *Br. Med. J.*, **292**, 1726–7.

• Blue blood

The hereditary possession of blue blood is traditionally the prerogative of royalty and the higher aristocracy. As W. S. Gilbert has reported, when speaking (or rather singing) through the agency of Lord Tolloller

> We boast an equal claim
> With him of humble name
> To be respected!
> Blue blood! Blue blood!
>
> (*Iolanthe*, Act I)

Certain hereditary disorders of the blood have affected the royal families of Europe, but haemophilia does not discolour the blood, and porphyria merely discolours the urine. Although the veins are printed a bright blue in anatomy textbooks, when medical students transfer to clinical studies they soon learn that an exposed vein is predominantly grey with a faint bluish tinge. Furthermore, whilst arterial blood is scarlet, venous blood is maroon.

The *OED* tells us that 'blue blood' is a translation from the Spanish *sangre azul*, and is 'that which flows in the veins of old and aristocratic families, who claimed never to have been contaminated by Moorish, Jewish, or other foreign admixture; the expression probably originated in the blueness of the veins of people of fair complexion as compared with those of dark skin'. This was eloquently evoked by Robert Browning, 'Blue as a vein o'er the Madonna's breast'.[1] Long before the blood groups and their genetics were known, blood was popularly treated as the typical component of the body which is inherited from parents and ancestors; hence blood in this sense denotes family, kindred or race.

The blueness of veins seen through translucent, lightly pigmented skin is partly due to scattering of light at the blue end of the visible spectrum when reflected against the dark veins beneath, much as the sky is blue against the blackness of outer space.

[1] Browning, R. (1845). 'The Bishop orders his tomb at St Praxed's Church'. In *Dramatic romances and lyrics*.

• Gullet

The Latin for throat, *gurgulio*, has some interesting derivatives. GORGE and GULLET are laymen's terms for pharynx and oesophagus. Gargoyles (Fr. *gargouille*) are rainwater spouts which project from roof gutters. On cathedrals they are sculpted as grotesque human or animal figures, a reaction, I suspect, to months of sculpting saints. GARGOYLISM (Hurler's syndrome) is one of the mucopolysaccharidoses, a group of inherited disorders based on an enzymic malfunction. It is so called because of the severe distortion of the face with which patients are affected, and which is reminiscent of architectural gargoyles.

GARGANTUAN, enormous, derives from Gargantua, a fictional giant created by a doctor turned novelist, François Rabelais (*c.*1455–1553). Enormous meals went down the throat of Gargantua, who was named after the Spanish for throat, *garganta*. JARGON, a private language, unintelligible to the outsider and used by a group, trade or profession, originally meant any noise made in the throat. This term cannot be applied to foreign languages, but the Danish wife of one of my colleagues admitted empathetically that Danish was 'not a language but a throat condition'.

To GARGLE is the act of rinsing the fauces with fluid in the mouth; the head is tilted backwards while breath is expelled to produce a bubbling sound, to wit, a GURGLE. Whereas it was formerly included in the routine of morning hygienic ablutions, the practice seems to be less prevalent nowadays, possibly as a result of increased faith in the efficacy of antibiotics. In its heyday it was usual to phonate during the expiratory phase, and a change of vowel sounds during the act was a refinement added, no doubt, to frighten the germs.

• Hair

Of the many Latin words for hair, two have medical connections. PILUS, now obsolete in English usage, remains with us in combining forms; PILOSEBACEOUS, PILOMOTOR,

PILONIDAL, possibly pilocarpine (Gk. *pilos*, wool; *karpos*, fruit). It is also buried within words, as in DEPILATE, DEPILATORY, and CAPILLARY (L. *caput*, head; *pilus*, hair of the head), whence slender and elongated like a scalp hair, when referring to a tube or blood vessel (see *Capillary*).

Shaggy animal hair is *villus* in Latin, whence we have VILLUS, one of the innumerable hair-like villi which give the intestinal mucosa a velvety (L. *villutus*) surface. VELLUS (L., fleece) is the fine body hair, excluding that which appears at puberty. VELOUR (F., velvet) is a term that has been adopted from drapery to describe a special weave of Dacron for use in arterial prostheses.[1]

[1] Kidson, I. G. (1983). Arterial prostheses. *Br. J. Hosp. Med.*, **30**, 248–54.

• Sphinx

The word SPHINX brings to the mind's eye of most readers that massive stone sculpture at Gizeh in Egypt, with a man's head and a lion's body. It is known locally as Harmakhis, and it represents royalty. Unconnected directly therewith is the Greek Sphinx. The Greeks named the Egyptian model 'Sphinx' through mistakenly identifying it with a similar monster of their own mythology. The Greek Sphinx had a woman's head, lion's body, serpent's tail, and eagle's wings; she was the issue of a union between the two-headed hound, Orthros, and his mother, Echidne (who also gave birth to the Chimaera, a fire-breathing goat with a lion's head and serpent's tail, the Hydra, the many-headed water serpent, and Cerberus, the three-headed Hound of Hell). One suspects that Echidne was exposed to teratogens, or maybe the Greeks anticipated genetic engineering. Now Hera, wife of top-god Zeus, had sent the Sphinx to punish

Thebes (never mind what for) by asking every Theban wayfarer a riddle. They all failed to give the right answer and were rewarded by being strangled and then devoured. But along came 'Swollen foot' (Oedi/pus) who solved the riddle, whereupon the Sphinx leapt from her perch on the mountain and was dashed to pieces in the valley below.

Sphinx means strangler (Gk. *sphingein*, to bind tightly, squeeze, throttle; *sphingion*, a necklace). The constrictive connotation gives us SPHINCTER, a circular muscle whose contraction partially or completely closes an orifice or the lumen of a tube.

The riddle of the Sphinx gives us SPHINGOSINE, the base of SPHINGOMYELIN and other SPHINGOLIPIDS, which are found in nervous tissue; also SPHINGOLIPIDOSIS, the collective name for a group of hereditary diseases in which an enzyme defect leads to the abnormal storage of sphingolipids in one or more tissues of the body, for example, Tay–Sachs disease. Johann Ludwig Wilhelm Thudichum (1829–1901) left his native Germany in 1854, and settled in England, where he became Director of the Chemistry and Pathology Laboratories at St. Thomas's Hospital. Describing his work on the chemistry of nervous tissue in 1881, he wrote, 'A body remained insoluble (in ether) . . . and to which, in commemoration of the many enigmas which it presented to the inquirer, I have given the name of *Sphingosin*'.

The Sphinx story is the earliest account of a viva voce examination and the dire consequences of failure. But whereas no examiner has leapt from a window on receiving the right answer, several have nearly died of boredom.

• Parotid, carotid

These words form a good rhyme, but are otherwise unrelated. PAROTID derives from Gk. *para*, beside; *ous*, gen. *ōtos*, ear, which aptly describes the parotid gland's anatomical position.

The usual adjectival suffix for ear is -otic, as in periotic (capsule), which it shares with words ending in -osis, as in cyanosis, cyanotic. I can think of no other ear-related words ending in -otid. The carotid arteries lie, at their cephalad ends, near the ears, but this is terminologically coincidental. The *OED* tells us that CAROTID stems from Gk. *karotides*, from *karoun*, 'to plunge into a deep sleep, to stupefy, because compression of these arteries is said to produce *carus* or stupor' (quoting Galen but not citing a specific reference). L. *carus* is a term applied to various forms of profound sleep or insensibility, especially 'the fourth or extremest degree of insensibility, the others being sopor, coma and lethargy'.[1] With the disappearance of carus from current usage, coma has now shifted its relative position to that of carus, whereas lethargy is now the mildest form, and about which I would submit for publication in the *British Medical Journal* a 'Personal View', if only I could find the energy to do so.

Of especial interest is the current use of the term 'sleeper' in wrestling for the manoeuvre of compression of a contestant's neck to cause loss of consciousness. Once again, 'the Greeks had a word for it'.

[1] Power, H. and Sedgwick, L. W. (1879–99). *New Sydenham Society Lexicon of medicine and applied science.* New Sydenham Society, London.

• Agony

AGONY in a medical context is severe, perhaps unbearable, pain. The *OED* says, '. . . so as to produce writhing or throes', but severe pain may also be associated with immobility. Agony is a word used by the laity rather than by doctors, and especially by patients who deem it advisable to convey the extent of their suffering to a doctor whom they suspect of being callous. In Greek, *agōnia* is a contest, a struggle for victory in games; from Gk. *agōn*, a sports gathering, where Gk. *agonistēs* is a contestant. The facial expression of a sprinter breasting the tape, or indeed of anyone engaged in intense adversarial physical effort, is indistinguishable from that of one who is in an agony of pain. It seems likely that this is the route whereby the transition from sport to pain has occurred.

Agōnia, in turn, derives from Gk. *agein*, to lead or drive, and is related to L. *agens*, present participle of *ago*, to put in motion (cf. Eng. agent), whence AGONIST is a muscle in a state of contraction, with reference to its opposing muscle, the ANTAGONIST, or a chemical substance capable of combining with receptors to initiate action, whence also AGONISM, to describe this phenomenon. An opponent in any sphere of human activity can be described as an antagonist and his attitude is said to be antagonistic, but curiously the reverse (agonistic, for helpfully friendly) does not exist. AGONAL describes a condition observed in a dying person, or one found in a body at necropsy, that occurred at the time of death. This term arose from the mistaken belief that the process of dying was necessarily painful. A PROTAGONIST was originally an actor taking the leading part, (Gk. *prōtos*, first; *agonistēs*, contestant) and now means the champion of a cause.

• The 'correct' spelling

From time to time my opinion is sought on the correct spelling of a word. This has always been presented as a choice of alternatives. I thought the following four examples might interest the reader. Uniformity is desirable but not always possible. Editors could, by agreement, play a useful role in spreading uniformity of practice.

Fetus versus foetus

Fetus is the correct Latin. But what is the correct English? We must first turn to the Spanish encyclopaedist, theologian, and historian, Archbishop Isidore of Seville (AD 560–636). His encyclopaedia contained extracts from the works of previous encyclopaedists; it attempted to comprise all contemporary knowledge, including etymology. It was he who introduced the spelling 'foetus', apparently in the belief that it could be derived from L. *foveo*, to warm, to cherish, and not from *feo*, beget.[1] 'Foetus' entered the English language, according to the *OED*, in 1594, since when this spelling has persisted in English writing. A minority of writers use 'fetus', and this fairly recent change appears to have emanated from America, and is gaining ground. 'Fetus' has the marginal advantage of being shorter. The oe (pronouned ee) of foetus gives it a spurious appearance of deriving from Greek. A lively and erudite discussion of the pros and cons appeared in the British Medical Journal in 1967.[1] One correspondent suggested abandoning both forms ('A plague o' both your houses!') and substituting embryo for all stages of intrauterine life. *Partus* is the commoner Latin for the young while still in the womb, according to Professor M. M. Willcock.

Certainly, foetus has been current in English for nearly four hundred years and is entitled to be regarded as naturalized. The fact that the Latin is *fetus* is no ground for change. What chaos would be created if the spelling of all English words reverted to their etymological roots! There is no need to indulge in heated controversy, nor to emulate the inhabitants of Lilliput and Blefuscu, who nearly came to war over the question, Should an egg be opened at the big end or the small end?[2] I regard both spellings as correct, and time will tell what direction the trend takes. I hope editors, meanwhile, will accept both spellings. We lack the equivalent of the *Académie Française*, so it will sort itself out in time.

[1] [Letters from various correspondents] (1967). *Brit. Med. J.* pp. 425, 568, 631.
[2] Swift, J. (1762). *Gulliver's travels.* (Reprinted 1967.) Penguin Books, Harmondsworth.

Calix versus calyx

Calix and calyx are not alternative spellings of one word; they are separate words. CALIX is derived from the Latin for cup which, in turn, comes from the Greek, *kylix* (Fig. 18). CALYX is from the Latin for any covering, husk, pod, shell, which derives, in its turn from the Greek, *kalyx*. Because of the similarity of the spellings, pronunciations and, to a lesser extent, meanings, confusion has arisen between them. The *OED* says on this, 'In mediaeval Latin and in the Romanic languages,

Fig. 18 Kylix—Attic red figure. (Courtesy of Sotheby & Co., London.)

this word [*calyx*] has run together in form with the much commoner Latin word *calix* "cup, goblet, drinking vessel"; and the two are to a great extent treated as one by modern scientific writers, so that the *calyx* of a flower and its derivatives are applied to many cup-like organs, which have nothing to do with the *calyx* of a flower, but are really meant to be compared to a *calix* or cup'. This commentary, though written in the period 1888–93, is valid today.

The earliest English use in the anatomical sense, quoted by the *OED*, is 'calyx' in Robert Knox's translation (1831) of J. H. Cloquet's *System of Human Anatomy* (1798). Other anatomists used other terms. James Keill, in *The Anatomy of the Human Body* (7th edn), 1723, describes the calices thus; 'The pelvis sends out several ramifications, which divide the urinary tubules into bundles, and which make a sort of *capsula* to the blood vessels'. Laurentius Heister, in the 1721 English translation of his *Compendium anatomicum*, writes of calices, 'The pelvis is a membranous cavity of the kidneys, emitting productions called *Tubuli Pelvis*, which surround the renal papillae'. Before this time anatomists evidently thought these circumpapillary extensions of the renal pelvis did not merit special description. Even Marcello Malpighi makes no mention thereof in his *De viscerum structura exercitatio anatomica* (1666; *de renibus*, pp. 71–100),[1] though it is clear from his meticulous description of the kidney that he cannot have failed to notice their presence. The earliest mention in an English dictionary that I have found is in Robert Hooper's *Lexicon Medicum*, 7th edn (1839), which gives both spellings, (a) calyx, under that entry, and (b) calix, under the entry 'kidney'.

The *OED* clearly prefers calix to calyx, and in view of its cup-like structure and function, this is etymologically correct. One might add that 'calyx' is correct botanical usage, and Malpighi adopts this spelling in his *Anatome plantarum* (*de floribus)* (1625).

Calix, cup, has yielded CHALICE, a goblet or communion cup. It is not without interest to see the muddle our Continental neighbours have got into over this. The German for chalice is *Kelch* (an initial c does not occur in Germanic words); however, the botanical calyx is *Blütenkelch* (flower-cup). The French use *calice* for chalice, calix (renal, correctly), and for the botanical calyx. The Italians, as the linguistic descendants of the Romans, might be expected to know better, but they use *calice* for both calyx and chalice.

[1] Hayman, J. M. (1925). Malpighi's Concerning the structure of the kidneys. *Ann. Med. Hist.*, 7, 242–63. (An English translation of the relevant section in Malpighi's book.)

Orchiectomy versus orchidectomy

The Greek for testicle is *orchis*; it has no d in any derivatives. The Latin for the plant whose tubers resemble this human organ in shape is *orchis*, derived from the Greek. In naming this botanical family *Orchideae*, Linnaeus (1751) wrongly assumed that the stem of the Latin *orchis* was *orchid-*. This in turn led to the adoption of orchid for the plant and its flower,[1] whence presumably ORCHIDECTOMY[2] acquired its intrusive d. There is no etymological justification for this.

On purely etymological grounds orchisectomy or orchectomy would be appropriate (compare orchitis). Although there is precedent for ORCHIECTOMY[3], I can think of no other instance where -ectomy is preceded by a vowel. It can be argued that the intrusive consonant makes for easier enunciation and perception of the spoken word by splitting what would otherwise be an awkward diphthong (much as the French insert an l in *si l'on*, allegedly for this reason). Thus the avoidance of two vowels together (in orchidectomy) is probably based on euphony. Of the various -ectomies, in some the terminal syllable is retained (pneumon-, gast(e)r-, splen-), while in others it is dropped (col-on, polyp-ous, nephr-os, hyster-a), but in each case the first moiety ends in a consonant. 'Orchisectomy' is an etymologically reasonable alternative, but a possible objection to it is that of fallaciously suggesting that -sect- implies cutting.

The intrusive d has existed in botanical terminology for over two hundred years, and in the English language for nearly one and a half centuries, and I feel might now reasonably be granted naturalization.

Incidentally, another tuber, the humble potato, has been likened to a testicle; the Spanish *turma* can mean either. English has not been immune to this double meaning, as when Max Miller,* at a Royal Command Variety Performance, referred to King Edward's.

[1] Lindley, J. (1846). *School botany; or the rudiments of botanical science*. Bradbury & Evans, London.

[2] Hospital Reports, Melbourne Hospital (1870). Operations by D. J. Thomas. *Austral. Med. Jo.*, **15**, 277–8.

[3] Gould, G. M. (1894). *An illustrated dictionary of medical, biological and allied sciences*. Baillière Tindall & Cox, London.

Trophic versus tropic

There is no connection between these two words though their similarity of spelling has led to some confusion. They have entirely different meanings and are of different origin.

TROPHIC (Gk. *trophikos*, from *trophē*, nourishment) means of or pertaining to nutrition; hence, the control of nutrition, size or activity of a part of the body by hormonal, vascular, or nerve supply. Accordingly, this root is used as a substantive suffix, as in atrophy, hypertrophy, and dystrophy, and adjectivally, as in corticotrophic.

TROPIC (Gk. *tropē*, turning) is not used in that sense as a substantive or adjective, but as a suffix, for example, phototropic. Heliotrope is the name of a plant whose flower turns to follow the diurnal passage of the sun; hence also the

* Max Miller was a stand-up comic who flourished in British music-hall in the mid-twentieth century. A 'King Edward' is a variety of potato.

purplish colour of this flower. The cyanosis of patients suffering from influenzal pneumonia during the 1918 pandemic was described as heliotrope (see cyan-, p. 11). In biology tropism is the movement of a portion of an organism towards (or away from in negative tropism) a source of light, heat or other stimulus. Not to be confused with taxis which is movement of the entire organism. The tendency to use -tropic when -trophic is intended should be resisted as it confuses their respective meanings.

There is some phonetic shift from the p to the f sound and vice versa within and between languages. Thus, the German *Affe*, *Neffe*, and *hoffen* correspond to the English ape, nephew, (from L. *nepos*) and to hope. According to Professor Sidney Allen (*Vox Graeca*, Cambridge University Press) the ancient Greek Φ, conventionally pronounced f, had a plosive component resembling p. In Hebrew the same letter of the alphabet (Pe) serves for the p and f sounds. A shift from p to f has occurred with the word trophy, originally a memorial of victory in war, when an enemy had been turned back, and put to flight.

• Medical and legal terminology—some differences

In their professional activities, doctors and lawyers share a small patch of common ground. Misunderstandings between the professions, which may arise despite the best of intentions, are especially likely to occur when the relationship is adversarial. The reasons for this are partly semantic and partly cultural, by which I mean the result of differences in traditional patterns of thinking. The law is man-made; its definitions are precise, and when a definition appears to be unclear or ambiguous, often as the result of the wording of a statute, a judge of the court will establish an interpretation. By contrast, the science and art of medicine are based on disorders of bodily or mental function which are incompletely understood. When in doubt opinions regarding diagnosis or prognosis tend, therefore, to be based on probabilities, which in turn rest on evidence which is often incomplete. Let us see how this affects the choice of words used by these two professions. I shall take two extreme examples in order to contrast the difference.

When wishing to opine regarding the diagnosis, trend, or prognosis of a disorder, a doctor may be able to quote figures of probability based on similar cases, but in the mundane situations of clinical practice figures are not usually available. The doctor is then obliged to draw on experience, which may be described as cumulative memory, often fallible, which is then modified by personal bias. There is now available a spectrum of terms from which to choose, as follows. It passes from 'almost certainly', through 'very probably' and 'probably', to 'commonly' and 'not uncommonly' (very popular), 'occasionally', 'uncommonly', and 'rarely' to 'hardly ever'; it carefully avoids the pitfalls of 'always' and 'never'.

Now, lawyers are trained in the techniques of closing loopholes, leaving no stones unturned, and exploring every avenue. This is obviously of great importance when, for example, framing a criminal charge or drafting a contract. But sometimes this obsession for ultimate precision goes to their heads, as in the example (Fig. 19) of a summons to attend an inquest which I received some years ago. The present-day wording is less pernickety, but this does not invalidate my theme.

We can ignore the superfluity of capitalized initials as a stylistic quirk, common in legal documents.

'to wit' A redundant archaism that prompts the reply, 'O Constable, let it be wist I wot well what thou wottest'.

'Hand and Seal' The 'hand' or signature is that of

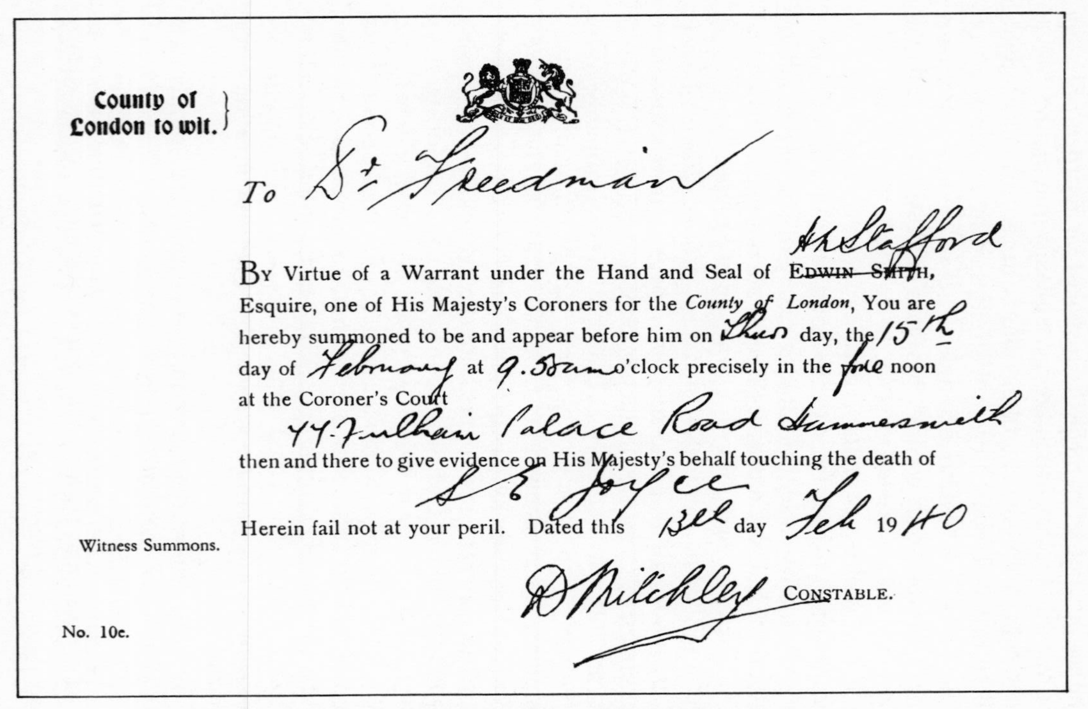

County of
London to wit.

To Dr Freedman

By Virtue of a Warrant under the Hand and Seal of ~~Edwin Smith~~ A H Stafford, Esquire, one of His Majesty's Coroners for the *County of London*, You are hereby summoned to be and appear before him on Thurs day, the 15th day of February at 9.30am o'clock precisely in the fore noon at the Coroner's Court
44 Fulham Palace Road Hammersmith
then and there to give evidence on His Majesty's behalf touching the death of
S E Joyce

Herein fail not at your peril. Dated this 3rd day Feb 1940

D Mitchley CONSTABLE.

Witness Summons.

No. 10e.

Fig. 19 Coroner's summons to give evidence at an inquest.

D. Mitchley, the Coroner's Officer; the seal was not applied to the document (perhaps the wax was in short supply).

'to be and appear' Yes, both simultaneously. Without the combined act, one might frustrate HM Coroner's summons by being without appearing. I could, for example, attend the Court boxed into a packing case, with holes for ventilation and the audibility of my voice. The reverse is also possible; I could appear without being before him by standing outside the courtroom window and giving evidence by hand signals. (I mean, you can't be too careful with these doctors.)

'then and there' Similarly, there must be a coincidence of time and place, otherwise I might choose to give evidence then but not there, by telephone or closed circuit TV (after all, the great auction houses accept bids of many thousands of pounds on the strength of such mediate communication). The reverse, 'there but not then', would tempt me perhaps to turn up when everyone else has gone to lunch, or at the right time on the wrong day.

'Herein fail not at your peril' This threat causes much disquiet. What horrid fate awaits me if I attend not 'precisely' but one minute early? 'We have spare room in the mortuary, Sir.' 'No, not that!' 'All right, seven days in custody for contempt of Court.' 'Thank you, that will be most acceptable after an 80-hour week on duty.'

'to give evidence on His Majesty's behalf' What exactly are they driving at? Is the King to give evidence? Apparently not; I am to give it on his behalf. Or is the coroner under the delusion that he is the King? Help, Kafka, let me out!

Nudgery

When the article 'Left and right' was printed in the *British Medical Journal*, I made a deliberate mistake in spelling the singular of species as specie. This rightly provoked a reader's response, to which I replied as follows.

The usage, meanings, and spellings of words in the English language have undergone innumerable changes in the last few centuries and, although these changes are slowing down, they still occur. Whenever a change occurs it is initiated by someone, somewhere. Every speaker and writer of the language can exercise the right to initiate or resist change. It seemed to me that, since the singular and plural of 'species' are identical, it would be useful to differentiate the one from the other. In offering for adoption 'specie' as the singular of species, my intention was to give the English language a nudge. The context would avoid confusion with specie meaning coined money.

I am not alone in nudgery. No less an authority than Professor Bernard Lennox has drawn attention to the desirability of exchanging Latin and Greek plurals for Anglicized ones.[1] Thus he proposes, for example, ganglions, criterions, and sulcuses. If *cheris* (compare F. *cerise*) can lose its terminal s to give cherry, so that we can now distinguish singular from plural, why not 'species'? The reverse process, that of plurals being treated as singulars, has become a commonplace, and has been swallowed without too much distaste.[2] For example, this *data* has been presented; the *agenda* has been circulated; the *media* is unfair to doctors; but the *news* is good.

I must now conclude as my *stamina* is not what it was.

[1] Lennox, B. (1980). *World Medicine* (23 Aug), 31.

[2] Howard, P. (1978). *Weasel words* (*Data is not what they used to be*). Hamish Hamilton, London.

General index

Index of foreign words

Latin verbs are shown in the first person singular present indicative; also in the infinitive, or the present or past participle, where any of these correspond to an English derivative. Nouns are shown in the nominative singular; also in the genitive, where this is relevant for the same reason. Latin terms for anatomical structures are in the General Index.

Glossary

aetiology: the study of fundamental causes.

areola: the pigmented disc of skin surrounding the nipple.

articular: pertaining to joints.

auscultation: the act of listening to sounds emanating from the interior of the body, normally performed with a stethoscope.

base (chemistry): a substance that reacts with an acid to form a salt.

bronchus: one of the air tubes in the lung, or of its branches.

cardiac: pertaining to the heart.

catabolism: the breakdown in the body of complex chemical compounds into simpler ones.

cilia: plural of cilium. Exceedingly small and closely packed hair-like projections from internal body surfaces. Their co-ordinated whiplike to and fro movements mediate the movement of particles along the surfaces.

coeliac disease: an inborn intolerance of the gluten (q.v.) in wheat and rye, causing diarrhoea and impaired absorption of dietary nutrients.

complement system: a group of serum proteins concerned with immunity. They are activated serially in a 'domino effect'.

cusp (in the heart): a thin membrane in a valve that flaps to and fro with each beat, thus ensuring one-way pumping of blood.

cyanosis: a bluish-mauve discoloration of those parts of the body (e.g. lips, cheeks, finger-nails) that are normally red. It is due to oxygen lack.

dextrocardia: exists when the heart is predominantly right-sided. May occur if the heart is displaced, and always present in situs inversus (q.v.).

DNA: deoxyribonucleic acid—the self-reproducing component of chromosomes and of many viruses, and the repository of hereditary characteristics.

duodenum: that part of the small bowel that extends from the outlet of the stomach.

encephalopathy: any disorder of brain function due to chemical or physical causes.

extra-corporeal: emanating from outside the body, and by implication affecting internal structures.

glycosides: organic compounds containing a sugar. In a medical context, they are the active principles of some drugs.

guanethidine: a drug used in the treatment of hypertension (q.v.).

hepatic: pertaining to the liver.

haemoglobin: a red substance present in blood corpuscles, and responsible for the colour of blood. It carries oxygen from the lungs to the body tissues.

haemolysis: destruction of red blood corpuscles.

halogen: one of a group of elements comprising fluorine, chlorine, bromine, and iodine.

hypertension: high blood pressure; an abnormally raised pressure of blood inside the arteries.

hypolactasia: a deficiency of the intestinal enzyme that digests lactose, the sugar in milk.

ileus: obstruction of intestinal movements due to mechanical causes or to paralysis of movement. Patients with ileus may become very distended with gas in the bowel.

interscapular: between the shoulder-blades.

ithyphallic: pertaining to the erect penis.

laevo-: an arrangement of atoms in a molecule that causes rotation of the plane of polarized light to the left, when the light passes through a solution of that substance.

lithotripsy: the breaking of bladder or kidney stones.

lumen: the hollow space inside any tubular structure, e.g. bowel, bronchus, blood vessel.

meatus, auditory: the ear-hole.

mitosis: the process of cell division involving a complex rearrangement of fibres (chromosomes) in the nucleus.

mitral stenosis: narrowing of the aperture regulated by one of the heart valves, called 'mitral' because of a fancied similarity in shape to a bishop's hat (mitre).

mucociliary clearance: the removal of unwanted particles embedded in a layer of mucus, by the action of cilia (q.v.).

otitis media: inflammation of the middle ear, namely the part that is situated on the inner side of the drum.

phenylketonuria: an inborn disorder of body chemistry that causes brain damage; treatable by special diet.

prosthesis: a fabricated substitute for a diseased or missing part of the body.

reniform: shaped like a kidney.

situs inversus: a condition in which all internal organs are reversed in mirror image situation from the normal; the heart is on the right; the liver, and appendix on the left.

sterno-clavicular joint: the joint connecting the collar-bone with breast-bone.

Tay–Sachs disease: an inborn disorder of body chemistry causing death at about one year of age.

temporo-mandibular joint: the joint connecting the jaw-bone with the base of the skull.

thalassaemia: an inherited disorder in which the haemoglobin (q.v.) molecule is abnormal, resulting in anaemia.

tracheostomy: cutting a hole surgically in the front of the windpipe to by-pass an obstruction to the airway.

uvula: the little dangler hanging from the back of the soft palate.

xanthines: a group of chemical substances. They are precursors of uric acid in normal body chemistry. In combination, they are components of the stimulants caffeine in tea, coffee, etc., and theobromine in chocolate.